透平机械：水轮机与水泵
专业英语

Professional English for Turbomachines:
Hydraulic Turbines and Pumps

王辉艳　刘小兵　主编

西南交通大学出版社
·成都·

图书在版编目（CIP）数据

透平机械：水轮机与水泵专业英语 = Professional English for Turbomachines: Hydraulic Turbines and Pumps / 王辉艳，刘小兵主编. —成都：西南交通大学出版社，2016.10
ISBN 978-7-5643-5048-2

Ⅰ.①透… Ⅱ.①王… ②刘… Ⅲ.①水轮机 – 英语 – 教材②水泵 – 英语 – 教材 Ⅳ.①TK73②TH38

中国版本图书馆 CIP 数据核字（2016）第 220628 号

透平机械：水轮机与水泵专业英语

Professional English for Turbomachines:
Hydraulic Turbines and Pumps

王辉艳　刘小兵　主编

责 任 编 辑	李　伟
助 理 编 辑	张文越
封 面 设 计	何东琳设计工作室
出 版 发 行	西南交通大学出版社 （四川省成都市二环路北一段 111 号 　西南交通大学创新大厦 21 楼）
发 行 部 电 话	028-87600564　028-87600533
邮 政 编 码	610031
网　　　　址	http://www.xnjdcbs.com
印　　　　刷	成都中铁二局永经堂印务有限责任公司
成 品 尺 寸	185 mm×260 mm
印　　　　张	20.5
字　　　　数	666 千
版　　　　次	2016 年 10 月第 1 版
印　　　　次	2016 年 10 月第 1 次
书　　　　号	ISBN 978-7-5643-5048-2
定　　　　价	58.00 元

课件咨询电话：028-87600533
图书如有印装质量问题　本社负责退换
版权所有　盗版必究　举报电话：028-87600562

前 言

本书可作为高等学校能源与动力工程专业水力机械及工程、水利水电动力工程方向本科生，以及动力机械及工程、流体机械及工程专业研究生的专业英语教材，还可以作为透平机械设计、制造、管理人员，水利水电工程设计、建设和管理人员专业英语参考书。

本书所选英文文献针对性较强，大多摘自原版英文技术资料、教材和相关网站，难度适中。本书共分为两大部分，第一部分主要讲述的是各种常用水轮机（切击式、斜击式、双击式、混流式、轴流定桨式、轴流转桨式、斜流式和灯泡贯流式）的结构、组成部分、各部分的功能、运行原理，以及水轮机的调速和选型；第二部分主要讲述的是离心泵的基本结构、基本参数、基本理论、能量损失、特性曲线、运行原理、空化空蚀，以及轴流泵和容积式泵。考虑到课堂教学的时间以及学生的知识结构等因素，本书共分为18章，每一章都是一个专题，每个章节的字数都在2 000英文单词左右，适合在两个学时内进行讲解。

为了方便学习，章节后附有本章中出现的重点词汇表，在书后还附有全部的英—中、中—英词汇对照表。每章后提出了总结性的问题，附有相关参考文献或参考网址，有助于教师、学生和专业人员查找相关的资料，并了解最新的动态。书后还附有每章内容的完整译文，供参考。

本书由流体及动力机械教育部重点实验室，四川省研究生教育教学改革项目，西华大学研究生教育教学改革项目资助出版，编者在此表示感谢。

鉴于编者水平有限，教材中难免存在不足与疏漏之处，敬请使用本书的师生和其他读者批评指正。

编 者
2016 年 5 月

CONTENTS

0 Exordium ··· 1
    0.1   Definition of a Turbomachine ·· 1
    0.2   Classification of Turbomachines ··· 1

1 Hydraulic Turbines ··· 4
    1.1   Introduction ··· 4
    1.2   Development History ··· 4
    1.3   Turbine Classification ··· 6
    1.4   Difference between Impulse and Reaction Turbines ···················· 11

2 Pelton Turbines ·· 15
    2.1   Injector ··· 15
    2.2   Servomotor ·· 16
    2.3   Deflector ··· 16
    2.4   Runner ·· 17
    2.5   Nozzle ·· 18
    2.6   Distributor ·· 18
    2.7   Casing ·· 19
    2.8   Braking Jet ·· 20

3 Turgo Turbines and Crossflow Turbines ·· 23
    3.1   Turgo Turbines ·· 23
    3.2   Crossflow Turbines ··· 24

4 Francis Turbines ·· 31
    4.1   Introduction ·· 31
    4.2   Main Components ·· 31

5 Propeller and Kaplan Turbines ·· 40
    5.1   Development ··· 40
    5.2   Main Components of the Runner ··· 41
    5.3   Scroll Case and Stay Ring ··· 44
    5.4   Guide Vane Cascade ·· 45
    5.5   Head Cover and Draft Tube ·· 45
    5.6   Automatic Air Valve ··· 46
    5.7   Shaft of the Hydro Unit ·· 46
    5.8   Over-speed Protective Device ··· 47

6 Deriaz Turbines ·· 50
    6.1   Introduction ·· 50
    6.2   Development ··· 51

		6.3	Mechanical Considerations	52
		6.4	Hydraulic Considerations	52
		6.5	Runaway Conditions	53
		6.6	Reversibility	53
		6.7	Servomotor	54

7 Bulb Turbines ... 56

- 7.1 Introduction ... 56
- 7.2 General Arrangement ... 56
- 7.3 Main Components ... 57
- 7.4 Assembly and Dismantling ... 66

8 Turbine Governing ... 69

- 8.1 Introduction ... 69
- 8.2 Turbine Governing Demands ... 69
- 8.3 Functions of Governors ... 70
- 8.4 Specific Turbine Governing Equipment ... 71

9 Selection of Hydraulic Turbines ... 76

- 9.1 Speed ... 77
- 9.2 Specific Speed ... 77
- 9.3 Turbine Efficiency ... 78
- 9.4 Turbine Setting and Excavation Requirements ... 80
- 9.5 Criteria for Selection of Hydraulic Turbines ... 82
- 9.6 Determination of the Number of Units ... 84

10 Classification and Components of Centrifugal Pumps ... 87

- 10.1 Classification by Flow ... 87
- 10.2 Stationary Components ... 88
- 10.3 Rotary Components ... 90
- 10.4 Shaft Seal Devices ... 91
- 10.5 Wearing Rings ... 92
- 10.6 Axial Thrust Balancing Devices ... 93
- 10.7 Multi-Stage Centrifugal Pumps ... 94

11 Basic Parameters of Centrifugal Pumps ... 97

- 11.1 Flow Rate ... 97
- 11.2 Head ... 97
- 11.3 Power ... 101
- 11.4 Efficiency ... 102
- 11.5 Rotational Speed ... 103
- 11.6 Net Positive Suction Head (*NPSH*) ... 103

12 Centrifugal Pump Theory ... 106

- 12.1 Velocity Triangles ... 106
- 12.2 Euler's Pump Equation ... 109

	12.3	Blade Shape and Pump Curve ································· 110
	12.4	Slip ··· 111
	12.5	Specific Speed of Centrifugal Pumps ······················· 113
13	**Pump Losses** ··· 116	
	13.1	Loss Types ··· 116
	13.2	Mechanical Losses ·· 117
	13.3	Hydraulic Losses ·· 117
	13.4	Loss Distribution as a Function of Specific Speed ····· 123
14	**Characteristics of Centrifugal Pumps and Their Systems** ···· 125	
	14.1	Pump Characteristic Curve ···································· 125
	14.2	System Characteristic Curve ·································· 128
	14.3	Matching of Pump and System Characteristics ········· 130
15	**Centrifugal Pumps Operating in Systems** ················ 132	
	15.1	Pumps Operating in Parallel ·································· 132
	15.2	Pumps Operating in Series ····································· 133
	15.3	Regulation of Pumps ··· 134
16	**Cavitation and *NPSH* in Centrifugal Pumps** ············ 142	
	16.1	Cavitation ·· 142
	16.2	Net Positive Suction Head ····································· 144
	16.3	Preventing Cavitation ··· 147
17	**Axial flow Pumps** ··· 150	
	17.1	Basic Structure of an Axial Flow Pump ··················· 150
	17.2	Blade Design ·· 152
	17.3	Airfoil Theory ·· 153
	17.4	Performance of an Axial Flow Pump ······················· 155
	17.5	Advantages of Axial Flow Pumps ···························· 156
	17.6	Applications of Axial Flow Pumps ·························· 157
18	**Positive Displacement Pumps** ································ 159	
	18.1	Introduction ·· 159
	18.2	Principle of Operation ·· 159
	18.3	Reciprocating Pumps ·· 160
	18.4	Rotary Pumps ·· 161
	18.5	Diaphragm Pumps ·· 165
	18.6	Positive Displacement Pump Characteristic Curves ··· 166
	18.7	Positive Displacement Pump Protection ·················· 166

参考译文 ·· 169

0	**绪论** ·· 169	
	0.1	透平机械定义 ·· 169
	0.2	透平机械分类 ·· 169

1 水轮机 ··· 171
   1.1 简介 ·· 171
   1.2 发展历史 ·· 171
   1.3 水轮机的分类 ·· 172
   1.4 冲击式水轮机与反击式水轮机的区别 ·· 177
2 切击式水轮机 ·· 178
   2.1 喷射机构 ·· 178
   2.2 喷针移动机构 ·· 179
   2.3 折向器 ··· 179
   2.4 转轮 ·· 179
   2.5 喷嘴 ·· 180
   2.6 配水管 ··· 180
   2.7 机壳 ·· 181
   2.8 喷射制动 ·· 182
3 斜击式水轮机和双击式水轮机 ·· 183
   3.1 斜击式水轮机 ·· 183
   3.2 双击式水轮机 ·· 184
4 混流式水轮机 ·· 188
   4.1 简介 ·· 188
   4.2 主要组成部件 ·· 188
5 轴流定桨和轴流转桨式水轮机 ·· 194
   5.1 开发 ·· 194
   5.2 转轮的主要组成部分 ··· 195
   5.3 蜗壳 ·· 198
   5.4 活动导叶栅 ··· 198
   5.5 顶盖和尾水管 ·· 198
   5.6 自动排气阀 ··· 199
   5.7 水电机组的主轴 ··· 199
   5.8 超速保护装置 ·· 199
6 斜流式水轮机 ·· 200
   6.1 简介 ·· 200
   6.2 开发 ·· 200
   6.3 机械特性 ·· 201
   6.4 水力特性 ·· 202
   6.5 失控工况 ·· 202
   6.6 可逆性 ··· 203
   6.7 接力器 ··· 203
7 灯泡贯流式水轮机 ··· 204
   7.1 简介 ·· 204

4

  7.2 总体布置 ··· 204
  7.3 主要组成部分 ··· 205
  7.4 组装和拆卸 ·· 212
8 **水轮机调速** ··· 213
  8.1 简　介 ··· 213
  8.2 水轮机调速的要求 ·· 213
  8.3 调速器的作用 ··· 214
  8.4 具体的水轮机调速设备 ··· 214
9 **水轮机选型** ··· 217
  9.1 转　速 ··· 218
  9.2 比转速 ··· 218
  9.3 水轮机效率 ·· 219
  9.4 水轮机安装高程和开挖要求 ······································ 220
  9.5 水轮机选型标准 ··· 222
  9.6 机组台数的确定 ··· 224
10 **离心泵的分类及组成部分** ··· 225
  10.1 按出流方向分类 ··· 225
  10.2 固定部件 ·· 226
  10.3 转动部件 ·· 227
  10.4 轴封装置 ·· 228
  10.5 承磨环 ·· 229
  10.6 轴向推力平衡装置 ·· 230
  10.7 多级离心泵 ·· 231
11 **离心泵的基本参数** ·· 232
  11.1 流　量 ·· 232
  11.2 扬　程 ·· 232
  11.3 功　率 ·· 234
  11.4 效　率 ·· 235
  11.5 转　速 ·· 236
  11.6 气蚀余量（NPSH） ·· 236
12 **离心泵基本理论** ·· 237
  12.1 速度三角形 ·· 237
  12.2 泵的欧拉方程 ··· 239
  12.3 叶片形状与泵的扬程曲线 ······································· 240
  12.4 反　旋 ·· 241
  12.5 离心泵的比转速 ··· 242
13 **泵的能量损失** ·· 244
  13.1 能量损失的类型 ··· 244

13.2 机械损失 ……………………………………………………………… 245
   13.3 水力损失 ……………………………………………………………… 245
   13.4 以比转速为函数的损失分布 ………………………………………… 249
14 离心泵及其系统的特性曲线 ……………………………………………… 250
   14.1 泵的特性曲线 ………………………………………………………… 250
   14.2 系统的特性曲线 ……………………………………………………… 253
   14.3 泵特性与系统特性的匹配 …………………………………………… 254
15 离心泵在系统中的运行 …………………………………………………… 255
   15.1 离心泵并联运行 ……………………………………………………… 255
   15.2 离心泵串联运行 ……………………………………………………… 256
   15.3 离心泵的工况调节 …………………………………………………… 257
16 离心泵的空化与气蚀余量 ………………………………………………… 261
   16.1 空化 …………………………………………………………………… 261
   16.2 气蚀余量 ……………………………………………………………… 262
   16.3 空化的预防 …………………………………………………………… 264
17 轴流泵 ……………………………………………………………………… 266
   17.1 轴流泵的基本构造 …………………………………………………… 266
   17.2 叶片设计 ……………………………………………………………… 267
   17.3 翼型理论 ……………………………………………………………… 268
   17.4 轴流泵的性能 ………………………………………………………… 270
   17.5 轴流泵的优点 ………………………………………………………… 271
   17.6 轴流泵的应用 ………………………………………………………… 271
18 容积式泵 …………………………………………………………………… 272
   18.1 简介 …………………………………………………………………… 272
   18.2 工作原理 ……………………………………………………………… 272
   18.3 往复泵 ………………………………………………………………… 272
   18.4 回转泵 ………………………………………………………………… 274
   18.5 隔膜泵 ………………………………………………………………… 277
   18.6 容积式泵的特性曲线 ………………………………………………… 277
   18.7 容积式泵的保护 ……………………………………………………… 278
Glossaries (English to Chinese) ………………………………………………… 279
Glossaries (Chinese to English) ………………………………………………… 298

0 Exordium

0.1 Definition of a Turbomachine

Turbomachines are defined as all those devices in which energy is transferred to or from a continuously flowing fluid by the dynamic action of one or more moving blade rows. The word "turbo" or "turbinis" is of Latin origin and implies that which spins or whirls around. Essentially, a rotating blade row, a rotor or an impeller changes the stagnation enthalpy of the fluid moving through it by doing either positive or negative work, depending upon the requirement of the machine. These enthalpy changes are intimately linked with the pressure changes occuring simultaneously in the fluid.

0.2 Classification of Turbomachines

0.2.1 According to Power Conversion Directions

Two main categories of turbomachines are identified: firstly, those absorb power to increase the fluid pressure or head (fans, compressors, and pumps); secondly, those produce power by expanding fluid to a lower pressure or head (wind, hydraulic, steam, and gas turbines). Fig.0.1 shows a variety of turbomachines commonly used in practice. The reason why so many different types of either pump (compressor) or turbine are in use lies in the almost infinite range of service requirements. Generally speaking, for a given set of operating requirements, one type of pump or turbine is best suited to provide optimum conditions of operation. Only hydraulic turbines and pumps are studied in this book.

0.2.2 According to Flow Paths

Turbomachines are further categorized according to the nature of the flow paths through the passages of the rotor. When the path of the through-flow is wholly or mainly parallel to the axis of rotation, the device is termed an axial flow turbomachine [e.g., Fig.0.1(a) and (e)].When the path of the through-flow is in a plane wholly or mainly perpendicular to the rotation axis, the device is termed a radial flow turbomachine [e.g., Fig.0.1(c)]. Mixed flow turbomachines are widely used. The term mixed flow in this context refers to the direction of the through-flow at the rotor outlet when both radial and axial velocity components are present in significant amounts. Fig.0.1(b) shows a mixed flow pump and Fig.0.1(d) shows a mixed flow hydraulic turbine.

0.2.3 According to Pressure Changes

All turbomachines can be classified as either impulse or reaction machines according to whether pressure changes are absent or present in the flow through the rotor. In an impulse machine all the pressure changes take place in one or more nozzles, the fluid being directed onto the rotor. The Pelton wheel, Fig.0.1(f), is an example of an impulse turbine. In a reaction machine, both the kinetic energy and pressure energy change when the fluid flows through the rotor. Fig.0.1(a-e) are all reaction machines.

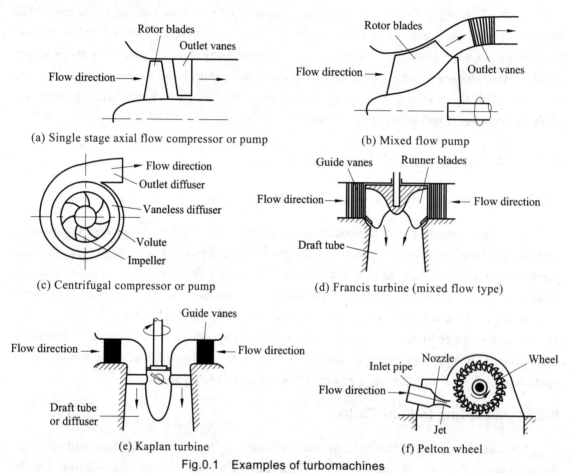

Fig.0.1 Examples of turbomachines

Questions

1. Define turbomachines.
2. Classify turbomachines according to power, flow path and pressure changes.

Reference

[1] S. L. Dixon and C. A. Hall. Fluid Mechanics and Thermodynamics of Turbomachinery [M]. 6th ed. Oxford: Butterworth-Heinemann, 2010.

Important Words and Phrases

absent 缺少的
axial flow 轴流；轴向流动
axis of rotation 旋转轴；转动轴
blade row 叶栅；叶片排
category 种类；分类
centrifugal compressor 离心式压缩机
compressor 压缩机
diffuser 扩散器；导轮
draft tube 尾水管
dynamic action 动力作用；动态作用
enthalpy 焓
exordium 绪论
expand 扩张；扩大
fan 风机
flow path 流动路径
fluid 液体；流体
Francis turbine 混流式水轮机
gas turbine 燃气轮机
guide vane 导叶
head 水头
hydraulic turbine 水轮机
identify 辨认；识别
impeller 叶轮
impulse turbine 冲击式水轮机
impulse 冲击式
infinite 无限的；无数的
jet 射流
Kaplan turbine 转桨式水轮机
kinetic energy 动能
mixed flow hydraulic turbine 混流式水轮机
mixed flow pump 混流泵
mixed flow 混流
negative work 负功
nozzle 喷嘴
optimum 最适宜的
outlet 出口
parallel 平行的
passage 通道
Pelton wheel 切击式水轮机
perpendicular 垂直的；正交的
plane 平面
positive work 正功
power conversion 能量转化
power 能量；动力
pressure energy 压能
pump 泵
radial flow 径向流；辐流
reaction 反击式
rotating 旋转；转动
rotation axis 旋转轴；转动轴
rotor 转子；转轮
simultaneously 同时地
spin 快速旋转
stagnation enthalpy 滞止焓
steam turbine 蒸汽轮机
through-flow 过流
turbo 涡轮
turbomachine 透平机械；透平机械；涡轮机
vane 叶片
vaneless 无叶片的
velocity component 速度分量
volute 蜗壳
wheel 转轮
whirl 旋转；回旋
wind turbine 风轮机

1 Hydraulic Turbines

1.1 Introduction

Hydraulic machine is a device in which mechanical energy is transferred from the liquid flowing through the machine to its operating member (runner, piston and others) or from the operating member of the machine to the liquid flowing through it. Hydraulic machines in which, the operating members receive energy from the liquid flowing through it and the inlet energy of the liquid is greater than the outlet energy of the liquid, are referred as hydraulic turbines. Hydraulic machines in which energy is transmitted from the working member to the flowing liquid and the energy of the liquid at the inlet of the hydraulic machine is less than the outlet energy are referred to as pumps.

It is well known from Newton's Law that to change the momentum of fluid, a force is required. Similarly, when the momentum of fluid is changed, a force is generated. This principle is also used in hydraulic turbine. In a turbine, blades or buckets are provided on a wheel and directed against water to alter the momentum of water. As the momentum is changed with the water passing through the wheel, the resulting force turns the shaft of the wheel performing work and generating power.

Hydraulic turbine uses potential energy and kinetic energy of water and converts it into usable mechanical energy. The mechanical energy available at the turbine shaft is used to run an electric power generator which is directly coupled to the turbine shaft. The electric power which is obtained from the hydraulic energy is known as hydroelectric energy.

1.2 Development History

The hydraulic turbine has a long period of development. Its oldest and simplest form is the water wheel, which was firstly used in ancient Greece and subsequently adopted throughout medieval Europe for the grinding of grain, etc. It was a French engineer, Benoit Fourneyron, who developed the first commercially successful hydraulic turbine (circa 1830). Later Fourneyron built turbines for industrial purposes that achieved a speed of 2300 r/min, developing about 50 kW at an efficiency of over 80 percent.

The American engineer James B. Francis designed the first radial-in-flow hydraulic turbine which became widely used, gave excellent results and was highly regarded. In its original form it was used for heads of between 10 m and 100 m. A simplified form of this turbine is shown in Fig.1.1. It can be observed that the flow path followed is essential from a radial direction to an axial direction.

Fig.1.1　A simplified Francis turbine

The Pelton wheel turbine, named after its American inventor Lester A. Pelton, was brought into use in the late nineteenth century. This is an impulse turbine in which water is piped at high pressure to a nozzle where it expands completely to atmospheric pressure. The emerging jet makes impacts onto the blades (buckets) of the turbine, producing the required torque and power output. A simplified diagram of a Pelton wheel turbine is shown in Fig.1.2. The head of water used originally was between about 90 m and 900 m (modern versions operate up to heads of 2,000 m).

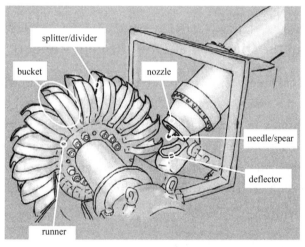

Fig.1.2　A simplified Pelton turbine

The increasing need for more power during the early twentieth century also led to the invention of a turbine suitable for small heads of water, i.e. 3 m to 9 m, in river locations where a dam could be built. It was in 1913 that Viktor Kaplan revealed his idea of the propeller (or Kaplan) turbine (see Fig.1.3), which acts like a ship's propeller in reverse rotation. Then Kaplan improved his turbine by means of swivelable blades, which improved the efficiency of the turbine in accordance with the prevailing conditions (i.e. the available flow rate and head).

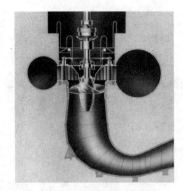

Fig.1.3 A simplified Kaplan turbine

1.3 Turbine Classification

The turbine has vanes, blades, or buckets that rotate around the axis by the action of the water. The rotating part of the turbine or water wheel is often referred to as the runner. The rotary action of the turbine in turn drives an electrical generator to produce electricity or drives other rotating machineries. Hydraulic turbines are machines that develop a torque from the dynamic and pressure action of water.

The hydraulic turbines are classified according to the flowing direction of water through vanes, the head at the inlet of the turbines, the specific speed of the turbines, and the pressure change of liquid through the runner.

1.3.1 According to the Direction of Flow through Runner

Tangential flow turbines: In this type of turbines, the water strikes the runner in the tangential direction to the wheel. Example: Pelton wheel turbine (Fig.1.2).

Radial flow turbines: In this type of turbines, the water flows in radial direction. The water may flow radially from outwards to inwards or vice versa. If the water flows from outwards to inwards through the runner, the turbine is known as inward radial flow turbine. Example: old Francis turbine. If the water flows from inwards to outwards, the turbine is known as outward radial flow turbine. Example: Fourneyron turbine (Fig.1.4).

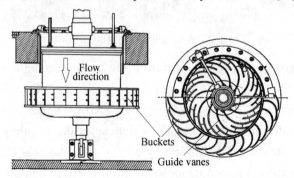

Fig.1.4 Fourneyron turbine

Axial flow turbines: The flow of water is in the direction parallel to the axis of the shaft. Example: Kaplan turbine and propeller turbine (Fig.1.3).

Mixed flow turbines: The water enters the runner in the radial direction and leaves in axial direction. Example: Modern Francis turbine (Fig.1.1).

1.3.2 According to the Head at the Inlet of Turbine

High head turbine: In this type of turbines, the net head varies from 150 m to 2000 m or even more, and these turbines require a small quantity of water. Example: Pelton wheel turbine.

Medium head turbine: The net head varies from 30 m to 150 m, and these turbines require moderate quantity of water. Example: Francis turbine.

Low head turbine: The net head is less than 30 m and these turbines require large quantity of water. Example: Kaplan turbine.

1.3.3 According to the Disposition of the Turbine Shaft

Turbine shaft can be either vertical or horizontal. In modern practice, Pelton turbines usually have horizontal shafts whereas the rest, especially the large units, have vertical shafts.

1.3.4 According to the Specific Speed of the Turbine

The specific speed of a turbine is defined as the rotating speed of a geometrically similar turbine that will develop unit power when it works under a unit head (1m head). It is prescribed by the relation $n_s = \dfrac{n\sqrt{P}}{H^{5/4}}$.

Low specific speed turbine: The specific speed is less than 50 (varying from 10 to 35 for single jet and up to 50 for double jet). Example: Pelton wheel turbine.

Medium specific turbine: The specific speed varies from 50 to 250. Example: Francis turbine.

High specific turbine: the specific speed is more than 250. Example: Kaplan turbine.

1.3.5 According to the Pressure Change of Liquid through the Runner

Based on the pressure change of liquid when it flows through the runner, turbines can be grouped into two types. One type is the impulse turbine, which utilizes the kinetic energy of a high-velocity jet of water to transform the water energy into mechanical energy. The second type is a reaction turbine, which develops power from the combined action of pressure energy and kinetic energy of the water.

1.3.5.1 Impulse Turbines

An impulse turbine is driven by high velocity jets of water from a nozzle directed on to buckets fixed on the periphery of the runner. The resulting impulse spins the turbine and leaves the fluid with diminished kinetic energy. There is no pressure change of the fluid on the turbine buckets, all the pressure drop takes place in the nozzles. Before reaching the turbine, the fluid's pressure head is changed to velocity head by accelerating the fluid with a nozzle. Impulse turbines do not require a pressure casement around the runner since the fluid jet is created by the nozzle prior to reaching the buckets on the runner. Newton's second law describes the energy transfer for impulse turbines. In essence, the impulse turbines only convert the kinetic energy of fluid, not the pressure. Within this major type, several sub-types of impulse turbines exist, each of which works in a slightly different way.

1. Pelton Turbine

Pelton turbines are designed with a number of cup-shaped components connected around the circumference of a runner that is in turn connected to a central hub (Fig.1.2). Nozzles are positioned all around the runner, and they inject water into these cups, which change the potential energy of the water into kinetic energy by pushing the turbine's wheel around.

2. Turgo Turbines

Turgo turbines are a variation of Pelton turbines. However, instead of full cups, Turgo turbines have only half cups around the runner (Fig.1.5). The presence of these half-cups allows the water to enter and exit the cups faster and in greater amounts than with Pelton turbines, thereby providing a much higher level of energy efficiency.

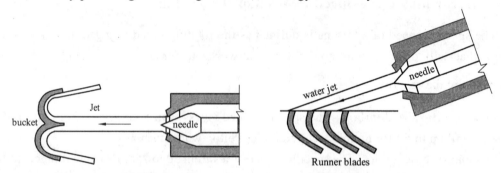

Fig.1.5 The blades and nozzles of a Pelton turbine and a Turgo turbine

3. Crossflow Water Turbines

Crossflow water turbines are designed with many trough-shaped blades in a radial arrangement around a cylinder-shaped runner (Fig.1.6). These blades are tapered at the water inlet as well as the blades' ends to ensure that the water flows as smoothly as possible. Crossflow turbines have only two nozzles, which shoot water at a 45-degree angle to the blades, thus converting the force into kinetic energy. A controlling mechanism regulates the flow of the water out of the nozzle. The water in these turbines actually passes through the blades twice, once from the outside of the blades to the inside, and another time from the inside to the outside. Crossflow turbines can usually handle a greater amount of water flow than Pelton turbines. They are also sometimes referred to as Michell-Banki or Ossberger turbines.

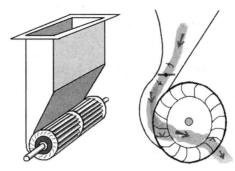

Fig.1.6 Crossflow turbine

1.3.5.2 Reaction Turbines

The reaction type of turbine works on the principle of reaction. Water enters the turbine at high pressure and low velocity in the guide passage. Some pressure energy is converted into kinetic energy. Water then enters the runner and pressure energy is successively converted into kinetic energy. As the water flowing through, the runner is accelerated, it creates a reaction on the runner vane and the runner is rotated.

As the static fluid pressure acts on both sides of the vane, it does not do any work. Work is entirely done due to the conversion of energy into kinetic form. It is to be noted that relative velocity goes on increasing from inlet to outlet though the absolute velocity keeps decreasing.

In a reaction turbine, turbine water is under pressure and the turbine is filled with water when it is working. Therefore, the turbine must be enclosed in a casing which should be able to withstand the pressure. In the case of an impulse turbine, the casing protects the runner and does not allow the water to splash out. It does not serve any hydraulic function.

In the operation of reaction turbines, the runner chamber is completely filled with water and a draft tube is used to recover as much hydraulic head as possible. Therefore, it is sometimes called a full admission turbine.

Water can pass through the hydraulic turbines in different flow paths. Based on the flow path of the liquid through the runner, hydraulic turbines can be categorized into three types: mixed flow turbines, axial flow turbines, and radial flow turbines.

1. Mixed Flow Turbines

For most of the hydraulic turbines used, there is a significant component of both axial and radial flows. Such types of hydraulic turbines are called as mixed flow turbines. Francis turbine is an example of mixed flow type. In a Francis turbine water enters in radial direction and exits in axial direction.

2. Axial Flow Turbines

Axial-flow turbines are generally either of the fixed-blade or Kaplan (adjustable-blade) variety. The "classical" propeller turbine is a vertical-axis machine with a scroll case and a radial wicket gate configuration that is very similar to the flow inlet for a Francis turbine, as shown in Fig.1.1 and Fig.1.3. The flow enters radially inward and makes a right-angle turn before entering the runner in an axial direction.

3. Radial Flow Turbines

Such hydraulic turbines have the liquid mainly flowing in a plane perpendicular to the axis of rotation. An early type of radial-flow machine was the Fourneyron turbine, in which water flow was radially outward.

Reaction turbines also include diagonal turbine or Deriaz turbine and tubular turbine.

4. Diagonal turbine or Deriaz turbine

Diagonal-flow turbines (Fig.1.7) bridge the gap between the propeller and the Francis turbines. The runner blades are set at an angle around the rim of a conical hub. There is no band around the blades. The blades are adjustable and can be feathered about an axis inclined at 45° to the axis of the shaft. The angle made by the blade axes with the main shaft decreases with increasing head and ranges between 30° and 60°.

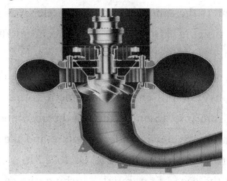

Fig.1.7 Deriaz turbine

5. Tubular turbine

The Tubular type turbines (Fig.1.8) are usually the best choices for exploitation of tidal power and hydraulic power with extremely low heads and extremely large flow rates. They have the advantages such as large discharge, high flow speed, high efficiency and less excavation, etc. The water diversion parts, runner and water drainage parts of tubular turbine are on the same axial line. The water flows directly and straightly through the runner without spiral casing. The water flow is axial from the pipe inlet to the draft tube outlet.

Tubular turbine generally has two categories: full tubular and half tubular. The generator's rotor of the full tubular type is on the outer circle of turbine's runner, whose application is less because of its difficult sealing. For half tubular turbines, the generator is separated from the hydro turbine. Their variations have Bulb, Pit and S types according to their structural types.

For bulb type tubular unit, the generator set is installed inside the airtight bulb body, which has a compact structure, straight passage shape, and high hydraulic efficiency thus it is widely used. For pit type tubular unit, the generator is installed inside a pit. For shaft extension type tubular unit, the generator is installed outside, the main shaft of turbine extends to the outside of the draft tube.

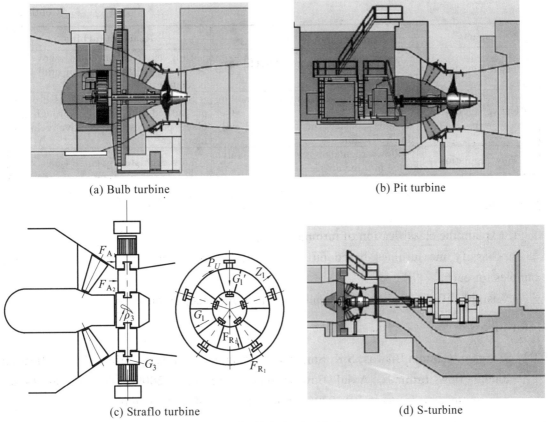

(a) Bulb turbine (b) Pit turbine

(c) Straflo turbine (d) S-turbine

Fig.1.8 Tubular turbines

1.4 Differences between Impulse and Reaction Turbines

The differences between Impulse and Reaction Turbines are listed in Table 1.1.

Table 1.1 Differences between Impulse and Reaction Turbines

Aspects	Impulse Turbine	Reaction turbine
Conversion of fluid energy	All the available energy of fluid is converted into kinetic energy by the nozzle	Only a portion of the fluid energy is converted into KE before fluid enters the runner
Changes in pressure and velocity	The pressure is constant (atmosphere) throughout the action of water on runner	Water enters the runner with an excessive pressure and both velocity and pressure change as water pass through runner
Action of water on blades	Blades are only in action when they are in front of the nozzle	Blades are in action all the time
Admittance of water over the wheel	Water may be allowed to enter a part or the whole of the wheel circumference	Water is admitted over the circumference of the wheel
Water tight casing	Water tight casing required	Not necessary
Extent to which water fills the wheel/turbine	Runner and blades are not completely filled or covered by water	Water completely fills all the passages between the blades throughout the operation of the turbine

Continue

Aspects	Impulse Turbine	Reaction turbine
Installation unit	Always installed above the tail race. No draft tube is used	Unit may be installed above or below the tail race – use of draft tube is required
Relative velocity of water	When water glides over the moving blades, its relative velocity either remains constant or reduces slightly due to friction	Since there is a continuous drop in pressure during flowing through the blade passages, the relative velocity increases
Flow regulation	Flow regulated by needle valve fitted into the nozzle	Flow regulated by guide vanes

Questions

1. Explain the classification of turbines.

2. Classify the turbines based on head, specific speed and hydraulic actions. Give examples for each.

3. List the differences between impulse turbines and reaction turbines.

Reference

[1] S. Sarkar, Gautam Biswas, S.K. Som and IIT Kanpur. Lecture 4 Turbines, Classification, Radial flow turbines, Axial flow turbines [DB/OL]. (2009-12-31) [2006-06-16]. http://nptel.ac.in/courses/105107059/module6/lecture4/lecture4.pdf.

Important Words and Expressions

accelerate 加速
adjustable-blade 调节叶片
airtight 密封的
atmospheric pressure 大气压
axial direction 轴向
axial flow turbine 轴流式水轮机
axial line 轴线
axis 轴
band 下环
Benoit Fourneyron 贝努瓦·富聂隆
blade axis 叶片轴
blade 叶片
bucket 斗叶；叶片
bulb type tubular unit 灯泡式贯流机组
bulb 灯泡；灯泡体
casement 机壳
casing 机壳

circa 大约；近似
circumference 圆周；周长
compact structure 结构紧凑；密实结构
conical hub 锥形轮毂
controlling mechanism 操作机构；控制机制
convert 转换
couple 耦合
crossflow water turbine 双击式水轮机
cylinder-shaped runner 圆柱形转轮
dam 大坝
deflector 偏流器
Deriaz turbine 斜流式水轮机
develop 产生；形成
diagonal turbine 斜流式水轮机
diminished 减少了的
discharge n. 流量；
　　　　　 v. 排放；流出；放出

disposition　排列；布置
divider　分水刃
drive　驱动
efficiency　效率
electric power　电力
enclose　围绕；装入
excavation　挖掘
feather　使……与……平行
fixed-blade　固定叶片
flow rate　流量
Francis turbine　弗朗西斯式水轮机
friction　摩擦
full admission turbine　整周进水式水轮机
fully tubular turbine　全贯流式水轮机
generator set　发电机组
generator　发电机
geometrically　几何学上地
glide over　滑过
grind　磨碎；碾碎
guide passage　引水道；引流部件
half tubular turbine　半贯流式水轮机
horizontal　水平的
hub　轮毂；中心
hydraulic efficiency　水力效率
hydraulic energy　水能
hydraulic head　水头；液压压头
hydraulic power　水力；水能；液压动力
hydroelectric energy　水电能源
in turn　依次；轮流地
inject　注入；注射
James B. Francis　詹姆斯·比切诺·法兰西斯
Kaplan turbine　卡普兰水轮机；转桨式水轮机
Lester A. Pelton　莱斯特·阿伦·佩尔顿
main shaft　主轴
mechanical energy　机械能
Michell-Banki turbine　双击式水轮机
mixed flow turbine　混流式水轮机
moderate　适度的；中等的；温和的

momentum　动量
needle　喷针
net head　净水头
Newton's Law　牛顿定律
Newton's second law　牛顿第二定律
operating member　运动部件
Ossberger turbine　双击式水轮机
parallel　平行
Pelton turbine　切击式水轮机；水斗式水轮机；佩尔顿水轮机
Pelton wheel turbine　佩尔顿式（切击式）水轮机
periphery　边缘；圆周；外围
piston　活塞
pit type tubular unit　竖井式贯流机组
pit　井；坑
potential energy　势能
power output　功率输出
prescribe　规定；表示
pressure action　压力作用
pressure drop　压力下降；压降
pressure head　压力水头；压头；压位差
prevailing　占优势的；主要的；普遍的；盛行的
propeller turbine　螺旋桨（式）涡轮机
radial direction　径向
radial flow turbine　径流式水轮机
radial-in-flow　径向流入式
reaction turbine　反击式水轮机
regulate　调节；规定；控制
resulting force　合力
reveal　显示；透露；揭露
reverse　反向
right-angle turn　直角转弯
rim　边；边缘
rotary action　旋转运动
rotate　旋转
rotating machinery　旋转机械
runner chamber　转轮室

runner 转轮
scroll case 蜗壳
sealing 封闭，密封
serve 具备
shaft extension type tubular unit 轴伸式贯流机组
shaft 轴
spear 喷杆
specific speed 比转速
splash out 飞溅出来
splitter 分水刃
strike 冲击；打击
S-type turbine 轴伸式水轮机
swivelable 可旋转的
tail race 尾水渠
tangent *adj.* 切线的；
　　　　 n. 切线，正切
tangential flow turbine 切向流式水轮机；冲击式水轮机
taper 使成锥形；逐渐减少
tidal power 潮汐能
torque 扭矩

trough-shaped blade 槽形叶片
tubular turbine 贯流式水轮机
Turgo turbine 斜击式水轮机；土戈尔式水轮机
unit head 单位水头
unit power 单位功率
variation 变化；变动；变异；变种
velocity head 速度头；动压头
velocity 流速；速度
vertical 垂直的
vertical-axis 竖轴；垂直式
Viktor Kaplan 维克多·卡普兰
water diversion part 引水部件
water drainage part 泄水部件
water energy 水能
water tight casing 水封的机壳
water wheel 水车
wheel 轮毂
wicket gate 活动导叶
work 功
working member 工作部件

2 Pelton Turbines

Pelton turbine is an impulse turbine wherein the flow is tangential to the runner and the available energy at the entrance is completely kinetic energy. It is preferred at a very high head and low discharges with low specific speeds. The pressure available at the inlet and the outlet is atmospheric.

All kinetic energy leaving the runner is "lost." A draft tube is generally not used since the runner operates under approximately atmospheric pressure and the head represented by the elevation of the unit above tailwater cannot be utilized. Since this is a high-head device, this loss in available head is relatively unimportant.

Pelton turbines can be arranged either by a horizontal or a vertical shaft. In general, a horizontal arrangement is found only in the medium and small sized turbines with one or two jets. Some horizontal Pelton turbines have however, been built with four jets as well. Large Pelton turbines with many jets are normally arranged with a vertical shaft. The jets are symmetrically distributed around the runner to balance the jet forces.

The Pelton turbine construction consists of a rather great number of components. Naturally some of these do not exist in every manufactured Pelton turbine. This matter of fact implies that the construction appears different from one manufacturer to the other or from a small to a big size of the turbine. For these reasons, only main components which are vital in all turbines are discussed.

2.1 Injector

Water is brought to the hydroelectric plant site through large penstocks, at the end of which there is an injector, which converts the pressure energy completely into kinetic energy, as shown in Fig.2.1. This will convert the liquid flow into a high-speed jet, which strikes the buckets or vanes mounted on the runner, which in-turn rotates the runner of the turbine. The amount of water striking the vanes is controlled by the forward and backward motion of the spear rod. As the water is flowing in the annular area between the nozzle opening and the spear, the flow gets reduced as the spear moves forward and vice-versa.

The functions of the injector are ① to direct the water received from the penstock at the proper angle on the runner ② to vary the quantity of water to suit instantaneous load conditions, thereby governing the turbine.

The injectors are located either at the end of a bend fitted to the penstock in the case of multi-nozzle turbines, or at the end of the distribution branches.

The injector consists of ① a nozzle ② a spear rod also called a needle and ③ a deflector. The spear rod slides coaxially in the nozzle. Its movement controls the area of the nozzle opening and therefore the quantity of water being admitted to the runner.

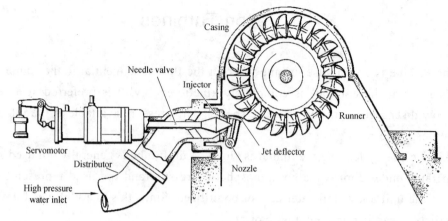

Fig.2.1 Cross section of a Pelton turbine

2.2 Servomotor

This has to be provided for movement of the spear. If the servomotor is outside the nozzle, a bend has to precede the straight portion of the nozzle to allow the spear stem to be brought out. In this type, the symmetry of the flow is only in one plane and the flow conditions inside the nozzle lead to a loss of energy. Though this type of nozzle gave reasonably good results for medium head medium output units, it is not suitable for modern high head, high output units. Also dismantling of the injector for repairs is difficult. Then a straight flow type injector (Fig.2.2) was developed, which keeps the water flow straight, by locating the servomotor inside the hub of the nozzle body. This is more efficient, smaller and of lighter construction and can be dismantled easily.

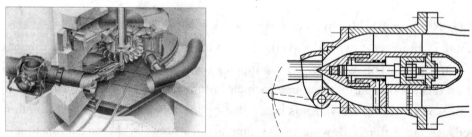

Fig.2.2 A straight flow type injector

2.3 Deflector

It is located at the exit of the nozzle preceding the runner buckets (Fig.2.2). Consequent upon a sudden rejection of load and a decrease in the demand of water, the needle quickly decreases the area of flow. Otherwise the turbine would start racing. But if the closing movement is too fast, the moving water column in the long penstock is suddenly decelerated, giving rise to a dangerous increase in pressure which may result in the bursting of the penstock. To achieve the twin objects of keeping the penstock pressure in limit and checking the racing tendency of the runner, the deflector is moved into the path of the jet partially or fully deflecting it from the runner. As the needle in the nozzle slowly closes and steady conditions are obtained, the deflector is moved back to its original position by means of a regulator. Its shape is carefully designed to cut off the jet to the needed extent. Care should be taken that the radial movement should not interfere with the moving buckets.

2.4 Runner

Runner is a cylindrical disk mounted on a shaft on the periphery of which a number of buckets are fixed equally spaced, as shown in Fig.2.3.

The blades, also called buckets, look like twin hemi-ellipsoided cups joined in the middle by means of a ridge. This shape is given to obtain maximum efficiency conditions. The jet of water enters the bucket in the center, bifurcates into two portions, travels over the bucket and leaves at the outlet tips. This bifurcation counter balances any axial thrust developed.

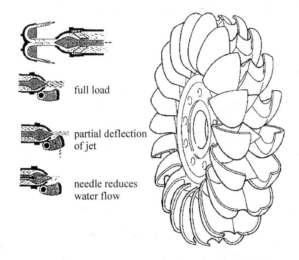

Fig.2.3 The runner of Pelton turbine

The jet is deflected at an angle of 168° to 176°, such that it just clears the back of the succeeding bucket.

For the most efficient conditions, the jet should hit the bucket perpendicularly. But this is not always possible as each bucket intercepts the jet at an angle of $2\pi/n$ (where n is number of buckets). For the jet to hit the bucket perpendicularly for the maximum time, the buckets are inclined backwards to the radius with the inclination varying from 10° to 18°.

A notch is provided at the tip of the bucket, so that the succeeding bucket does not interfere with the jet hitting the previous bucket.

The number of bucket is decided by two conditions:

① The number must be adequate to intercept the jet at all times and no water should go to the tail race directly without imparting energy to the runner.

② The number of buckets provided currently varies from 18 to 24, though the extreme limit is from 14 to 30.

2.5 Nozzle

The Pelton turbine can have either a horizontal axis or a vertical axis layout. If the axis is horizontal, the runner can be fed either by one nozzle or two nozzles. Sometimes the turbine may have twin runners placed side by side or alternatively, the generator can be driven by two turbines placed on either side.

In the vertical shaft disposition, there is only one runner. The number of nozzles varies from 2 to 6, and most of the turbines have either 4 or 6 nozzles.

The Pelton wheel is a low specific speed device. Specific speed can be increased by the addition of extra nozzles and the specific speed increases by the square root of the number of nozzles. However, it is a serious demand that the jets enter the buckets so far from each other that no mutual disturbances can occur. In practice this means that six jets represent the upper limit of their number.

2.6 Distributor

To feed water to these nozzles, a distributor is provided which is a pipe attached to the end of penstock. For a two-nozzle horizontal shaft layout, the distributor is Y-shaped with bends being attached to the two arms of the bifurcate for mounting the injectors (Fig.2.4). The angle of the Y varies between 70° to 90°.

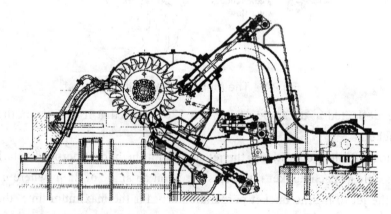

Fig.2.4 A two-nozzle horizontal shaft layout with Y-shaped distributor

For the vertical shaft layout, the simplest type of distributor consists of annular pipe starting from the penstock with successively decreasing cross-sectional area and with a runner of arms projecting inside for fitting of the nozzles (Fig.2.5).

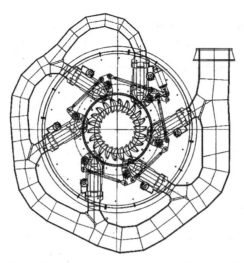

Fig.2.5　Distributor of a six nozzle Pelton turbine

2.7　Casing

The functions of the casing are:
① To enclose the turbine runner and jets so that the water does not spill;
② To provide recesses for mounting the injectors;
③ To provide recesses for mounting the bearing;
④ To transmit the load of the machine to the foundations;
⑤ To serve as a turbine pit liner;
⑥ To discharge the water into tailrace.

As the turbine operates at atmospheric pressure, the casing does not serve any hydraulic function.

For the horizontal shaft machine (Fig.2.6), the casing consists of a rectangular box at the bottom and a semicircular hood at the top with an intermediate transitory section. The lower section serves as a robust lining to protect the surrounding concrete. It is made of a welded steel plate provided with stiffeners and ties to obtain a perfect adhesion between the concrete and casing. An access door is provided for inspection.

A cast steel shield is provided to protect the housing from the jet in case of sudden runaway conditions. Baffles are provided to destroy the energy of any jet deflected by sudden movements of the governor. The transitory section starts from the top of the rectangular box and ends at the center

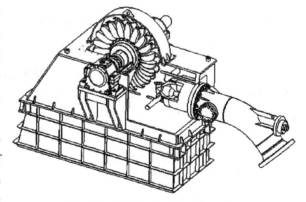

Fig.2.6　Pelton turbine's casing

of the turbine axis. Recesses on the transitory section are provided for accommodating the turbine bearing. The top semicircular hood is split at the axis and the runner is exposed as soon as it is removed. The hood is generally dimensioned closely to minimize the windage losses of the runner.

In case of two injector machines, an oblique division of the semicircular hood is provided which helps to remove the runner without dismantling the upper injector.

In the case of a vertical shaft turbine, the casing may be conical with the diameter increasing downwards, or cylindrical, or the top portion being conical and the bottom portion being cylindrical. The casing is fabricated from steel plates. The strengthening ribs and spikes are provided on the outside of the casing to provide a firm grip on the surrounding concrete. Openings have to be provided on the sides for the distribution pipes to be led inside.

2.8 Braking Jet

Even after the amount of water striking the buckets is completely stopped, the runner goes on rotating for a very long time due to inertia. To stop the runner in a short time, a small nozzle is provided which directs the jet of water on the back of bucket with which the rotation of the runner is reversed, as shown in Fig.2.7. This jet is called as braking jet.

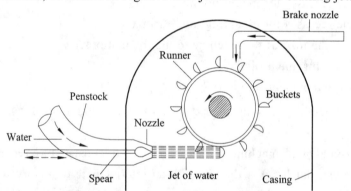

Fig.2.7 Schematic view of jet brake

The small nozzle receives water directly from the distributor and the nozzle is opened automatically when needed and automatically shut off to avoid the runner rotation in the opposite direction when the runner stops moving.

Questions

1. With a neat sketch explain the parts of a Pelton turbine.
2. State the purposes of the following Pelton turbine components:
   - Injector
   - Servomotor
   - Deflector
   - Casing

· Braking jet

3. Describe the characteristics of a Pelton turbine runner.

References

[1] Magdy Abou Rayan, Nabil H. Mostafa, Purage Ohans. Textbook of Machines Hydraulic [M]. Zagazig: Zagazig University, Dept. of Mechanical Engineering, 2009.

[2] Arne Kjølle. Hydropower in Norway-Mechanical Equipment [M]. Trondheim: Norwegian University of Science and Technology, 2001.09.

[3] M.S.Govinde Gowda. Hydraulic Turbines [EB/OL]. [2016-06-16]. http://elearning.vtu.ac.in/newvtuelc/courses/06ME55.html.

[4] Ingram G. Basic concepts in turbomachinery[M]. Frederiksberg: Grant Ingram & Ventus Publishing ApS, 2009.

[5] Voith Hydro. Erection of Pelton Turbine [EB/OL]. (2009-10) [2016-06-16]. http://www.voith.com/ca-en/Voith_Pelton_turbines.pdf

Important Words and Phrases

access door 人孔门；检修门；通道门
accommodate 容纳；使适应；供应；调解
adhesion 黏附；黏附力
annular area 环形区域
annular pipe 环形管
area of flow 过流面积
arm 支臂
available head 有效水头
axial thrust 轴向推力
baffle 挡板
bearing 轴承
bend 弯管
bifurcate 分叉
braking jet 制动射流
burst 破裂
cast steel 铸钢
check 制止；抑制
clear 避开；绕开
coaxially 同轴地
concrete 混凝土
conical 圆锥的；圆锥形的
construction 构造；结构
counter balance 平衡；配衡

cross section 横截面
cut off 切断
cylindrical disc 圆柱形的轮辐
cylindrical 圆柱体的
decelerate 使减速
deflect 使转向；使偏斜；使弯曲
deflection 偏向；偏转
deflector 折向器
demand of water 需水量
dimension 把……刨成（或削成）所需尺寸
dismantle 拆开；拆卸
distribution branch 配水支管
distribution pipe 配水管
distributor 进水管；配水管
elevation 高程；高度；海拔
expose 使暴露；使显露
fabricate 制造；装配
feed 向……提供；流入；注入；喂养
foundation 基础；地基
govern 调节
governor 调速器
grip 抓力；紧握
hemi-ellipsoided cup 半椭圆的杯子

horizontal axis 水平轴；横轴；卧轴	regulator 调节器；调节装置；调控机构
horizontal shaft 水平轴；卧轴	rejection of load 甩负荷
housing 机壳；外罩	remove 移除；移动；迁移
impart 传递	reverse 倒转；反转
inclination 倾角	robust 结实的；耐用的；坚固的
incline 使倾斜；使倾向于	runaway 失去控制的
inertia 惯性；惰性	semicircular hood 半圆形罩
injector 喷射机构；喷管	servomotor 接力器；移动机构
inspection 检查；检验	shield 盾；护罩；防护物
instantaneous load condition 瞬时负荷条件	spear rod 针阀
intercept 拦截；拦住	spear stem 喷杆
interfere 干涉；妨碍	spill 溢出；溅出
intermediate transitory section 中间过渡段	split 分开；分离；劈开
jet force 水冲力；喷射力	square root 平方根
layout 布置方式；布局；安排；设计	steady condition 稳定工况
lining 衬里；内层；衬套	steel plate 钢板
load 负载；负荷	stiffener 加固物；加劲杆；加强筋
loss of energy 能量损失	straight flow type injector 直流式喷射机构；直喷嘴；内控式喷射机构
manufacturer 制造商	
mount 安装	strengthening rib 加强肋
multi-nozzle turbine 多喷嘴水轮机	strengthening spike 加强钉
notch 凹口；凹槽；缺口	succeeding 随后的
oblique division 倾斜的部分	symmetrical 对称的；匀称的
opening 开度	symmetry 对称（性）
outlet tip 出水边	tailrace 尾水渠
output 出力	tailwater 尾水；下游水
partially 部分地	tangential 切向的；切线的；正切的
penstock 压力水管	tie 加强带；束缚
pit liner 基坑里衬	transmit 传输；传送；传递
precede 在……之前	unit 机组
previous 先前的；稍前的	utilize 利用
project 伸出；投影	vertical axis 垂直轴；纵轴；立轴
racing 飞逸	vertical shaft 垂直轴；立轴
radius 半径	water column 水柱
recess 凹槽；凹处	weld 焊接
rectangular box 矩形盒子；矩形底座；矩形框架	windage loss 风阻损失

3 Turgo Turbines and Crossflow Turbines

3.1 Turgo Turbines

The Turgo turbine is an impulse water turbine designed for medium head applications. It is a modification of Pelton wheel where the runner is a cast wheel whose shape generally resembles a fan blade that is closed on the outer edges, as shown in Fig.3.1. Operational Turgo turbines achieve efficiencies of about 87%. In factory and lab tests, Turgo turbines perform with efficiencies of up to 90%. It works with net heads between 15 and 300 m.

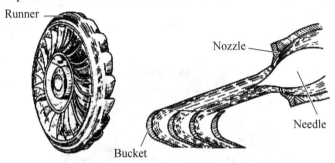

Fig.3.1　A Turgo impulse turbine

3.1.1　Advantages

Developed in 1919 by Gilbert Gilkes as a modification of the Pelton wheel, the Turgo has some advantages over Francis and Pelton designs for certain applications.

Firstly, the runner is cheaper to make than a Pelton wheel. Secondly, it doesn't need an airtight housing like the Francis turbine. Thirdly, it has higher specific speed and can handle a greater flow than the same diameter Pelton wheel, leading to reduced generator and installation cost. Turgo turbine is not only characterized by high specific speed, high flow and even efficiency curve, but also by simple construction, easy erection test and adjustment, reliable operation, simple maintenance and low cost.

Turgos operate in a head range where the Francis and Pelton overlap. While many large Turgo installations exist, they are also popular for small hydro where low cost is very important.

3.1.2　Theory of Operation

The Turgo turbine is an impulse turbine design which uses a special nozzle at the end of a pipe to convert the flow of the water into a high pressure jet. This jet of water is then directed at an angle of about 20 degrees to the face of the runner which uses spoon-shaped blades to capture the jet of water. A photo of a water jet impinging on the rotor of a Turgo turbine is shown in Fig.3.2. These blades are specially shaped so that the pressurized water enters the blades on one side, and then goes across the blade and exits on the other side, which convert the kinetic energy of the water jet into rotational shaft power. A Turgo runner looks like a Pelton runner split in half. For the same power, the Turgo runner is one half the diameter of the Pelton runner, so the specific speed doubles.

This tangential flow of the water across the turbines runner means that the waste water leaves the blades without interfering with the incoming water jet as is the case with other impulse turbines. The result is that a Turgo turbine can have a smaller diameter runner and blades for an equivalent power input allowing it to rotate at much higher speeds to drive an electrical generator.

One of the disadvantage of Turgo turbine designs is that to generate sufficient nozzle pressure to rotate the turbine runner at high speeds, they require more head height than other designs.

Fig.3.2 Water jet impinging on the rotor of a Turgo turbine

However, to overcome these limitations, instead of using just one nozzle and water jet to rotate the turbine runner, the Turgo turbine can be fitted with multiple nozzles located equally around its circumference. Then a Turgo turbine can be equipped with two, four or six jet nozzles placed around the turbine runner and blades to extract more kinetic energy from the water.

This increase in the number of water jet nozzles means that the Turgo turbine can operate under low-head and low-flow conditions. The amount of energy extracted from the water will increase as the power output is proportional to the number of nozzles.

Turgo turbine designs represent an improvement in the available shaft power compared to conventional water wheels and other impulse turbine designs. They are generally less expensive to make and with multiple nozzles they can handle a greater flow of water, so a smaller Turgo turbine can generate the same amount of power as a larger water wheel.

3.2 Crossflow Turbines

A crossflow turbine, Bánki-Michell turbine, or Ossberger turbine is a water turbine developed by the Australian Anthony Michell, the Hungarian Donát Bánki and the German Fritz Ossberger. Michell obtained patents for his turbine design in 1903. Professor Banki, a Hungarian engineer, developed the machine further. Ossberger brought this kind of turbine construction to production stage. Its development was patented first in 1933. Today, the company founded by Ossberger is the leading manufacturer of this type of turbine.

Unlike most water turbines with axial or radial flows, the water passes through the turbine transversely, or across the turbine blades in a crossflow turbine. The water flows firstly from the outside of the turbine to its inside, and then after passing the runner, it leaves on the opposite side, as shown in Fig.3.3. Going through the runner twice provides additional efficiency, but most of the power is transferred on the first pass and only 1/3 of the power is transferred to the runner when the water is leaving the turbine. When the water leaves the runner, it also helps clean small debris and pollution in the runner. The crossflow turbine is a low-speed machine that is well suited for locations with a low head but a high flow.

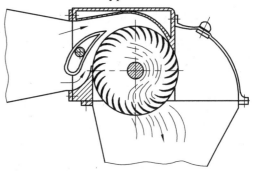

Fig.3.3   Schematic view of a crossflow turbine

Although the illustration shows one nozzle for simplicity, most practical crossflow turbines have two nozzles arranged so that the water flows do not interfere.

If the water flow is variable, the crossflow turbine is designed with two cells of different capacity that share the same shaft, as shown in Fig.3.4. The turbine wheels have the same

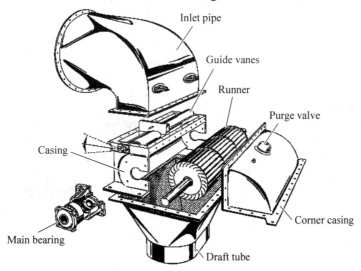

Fig.3.4   Ossberger turbine section

diameter, but different lengths to handle different volumes at the same pressure. The subdivided wheels are usually built with volumes in a ratio of 1 : 2. The narrower cell processes small water flow whereas the wider cell processes medium flow. Both cells process full flow together. The subdivided regulating unit, the guide vane system in the turbine's upstream section, provides flexible operation with 33, 66 or 100% output, depending on the flow. Low operating cost is obtained with the turbine's relatively simple construction.

3.2.1 Efficiency

The peak efficiency of a crossflow turbine is somewhat less than a Kaplan, Francis or Pelton turbine. The total efficiency of small crossflow turbines with a small head is between 80%-84% throughout the flow. The maximum efficiency of medium and big turbines with a higher head is 87%. However, the crossflow turbine has a flat efficiency curve under varying loads. With a split runner and turbine chamber, the turbine maintains its efficiency while the flow and load vary from 1/6 to the maximum. The advantages of a partially loaded crossflow turbine are illustrated by the efficiency curve shown in Fig.3.5.

Particularly with small run-of-the-river plants, the flat efficiency curve yields better annual performance than other turbine systems, as small rivers' water is usually lower in some months. The efficiency of a turbine determines whether electricity is produced during the periods when rivers have low flows. If the turbines used have high peak efficiencies but behave poorly at partial load, less annual performance will be obtained than the turbines that have a flat efficiency curve.

3.2.2 Turbine Casing

The casing of a crossflow turbine is made of structural steel; it is robust, resistant to impacts and frost. If there is high contents of abrasive material in the water (e.g. sand, silt) or if the actual composition of the water is considered aggressive (e.g. sea water, acidy water), all parts of the turbine in contact with water are made of stainless steel.

3.2.3 Guide Vanes

In a split crossflow turbine, the working water is directed by two force-balanced profiled guide vanes. The vanes divide the

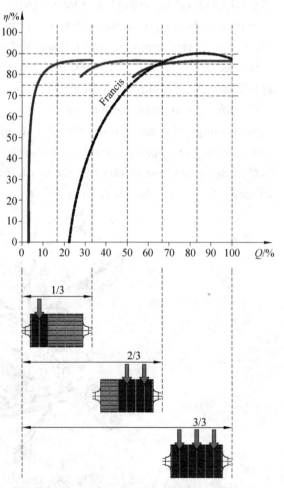

Fig.3.5 Crossflow turbine curve of efficiency (the flow is regulated by guide vanes in the proportion of 1 : 2) compared to a Francis turbine

water flow, direct it and allow it to enter the runner smoothly for opening with arbitrary width. Both rotary guide vanes are fitted very precisely into the turbine casing and they can serve as the closing device of the turbine if there is a lower head. The guide vanes therefore act as the valves between the penstock and turbine. Then it is not necessary to use a shutoff valve. Both guide vanes can be adjusted independently of one another via regulating levers to which the automatic or manual control is connected. Guide vanes are placed in highly resistant slide bearings, which do not require any maintenance. The turbine will be able to close by gravitation in the event of shutdown by adding weights to the arms ends.

3.2.4 Runner

The runner is the most important part of the turbine. The crossflow turbine has a cylindrical runner with a horizontal shaft. It is equipped with blades manufactured of bright-rolled profiled steel. The blade's edges are sharpened to reduce resistance to the flow of water. A blade is made in a part-circular cross-section (as pipes cut over its whole length). The ends of the blades are welded to disks, which forms a cage like a hamster cage and are sometimes called "squirrel cage turbines"; instead of the bars, the turbine has the trough-shaped steel blades, as shown in Fig.3.6. Depending on the size, the runner has up to 37 blades. The linearly slanted blades only create slight axial force thus reinforced axial bearings with complex fitting and lubrication are not required. Blades of wider runners are supported by multiple discs.

Fig.3.6 Schematic view of a crossflow turbine runner

3.2.5 Bearings

Crossflow turbines are equipped with self-aligning roller bearings which have several advantages, such as low rolling resistance and simple maintenance. The design of the bearing housing prevents water leakage into bearings and contact of lubricants with working water. Furthermore, the runner is centered in the turbine casing by means of the bearings. Apart from grease changes every year, the bearings do not require any maintenance.

3.2.6 Draft Tube

The crossflow turbine is a free stream turbine, just like a Pelton turbine. However, in case of medium or low head, it is possible to apply a draft tube in order to utilize the entire head. The water column in the draft tube must be controllable though. This is ensured by a regulation air valve, affecting the suction pressure in the turbine casing. In such a manner, turbines with a suction head from 1 to 3 m can be optimally used without any danger of cavitation.

3.2.7 Advantages

Since it has a low price and a good regulation ability, crossflow turbines are mostly used in mini and micro hydropower units of less than two thousand kW and with heads less than 200 m.

Due to its excellent behavior with partial loads, the crossflow turbine is well-suited to unattended electricity production. Its simple construction makes it easier to maintain than other turbine types; only two bearings must be maintained, and there are only three rotating elements. The mechanical system is simple, so repairs can be performed by local mechanics.

Another advantage is that it can often clean itself. As the water leaves the runner, leaves, grass etc. will not remain in the runner, preventing losses. Therefore, although the turbine's efficiency is somewhat lower, it is more reliable than other types. No runner cleaning is normally necessary, e.g. by flow inversion or variations of the speed. Other turbine types are clogged more easily, and consequently face higher power losses despite higher nominal efficiencies.

Questions

1. Compare the runners of Pelton turbines and Turgo turbines.
2. List the advantages of Turgo turbines over Pelton turbines.
3. Given a drawing of a crossflow turbine, identify the following major components:
   · Turbine casing
   · Guide vanes
   · Runner
   · Draft tube
   · Main bearing
4. Explain why the overall efficiency of a crossflow turbine is higher than a Francis turbine.
5. List the advantages of crossflow turbines.
6. Explain the operating principle of crossflow turbines.

References

[1] Magdy Abou Rayan, Nabil H. Mostafa, Purage Ohans. Textbook of Machines Hydraulic [M]. Zagazig: Zagazig University, Dept. of Mechanical Engineering, 2009.
[2] Arne Kjølle. Hydropower in Norway-Mechanical Equipment [M]. Trondheim: Norwegian University of Science and Technology, 2001.
[3] Renewables First. Pelton and Turgo Turbines [EB/OL]. [2016-06-16]. https://www.renewablesfirst.co.uk/hydropower/hydropower-learning-centre/pelton-and-turgo-turbines/.
[4] M.S.Govinde Gowda. Hydraulic Turbines [EB/OL]. [2016-06-16]. http://elearning.vtu.ac.in/newvtuelc/courses/06ME55.html.

[5] Ingram G. Basic concepts in turbomachinery[M]. Frederiksberg: Grant Ingram & Ventus Publishing ApS, 2009.

[6] CINK Hydro–Energy. 2-cell Crossflow Turbine [EB/OL]. [2016-06-16]. http://cink-hydro-energy.com/en/2-cell-crossflow-turbine.

[7] Cálculo de Preços. OSSBERGER-Turbine [EB/OL]. [2016-06-16]. http://www.ossberger.de/cms/pt/hydro/ossberger-turbine/.

Important Words and Phrases

abrasive material　磨蚀物质；研磨材料
acidy water　酸性水
across　穿过；横穿
aggressive　有腐蚀性；侵略性的；好斗的
air valve　空气阀
axial bearing　止推轴承
axial force　轴向力
Bánki-Michell turbine　双击式水轮机
bearing housing　轴承箱；轴承体
behave　表现；运转
bright-rolled　精轧的
capacity　流量；容量
capture　捕获；捕捉
cast wheel　铸轮
cavitation　空化；气穴现象；空穴作用
clog　阻碍；堵塞
closing device　关闭装置；闭合装置
composition　构成；成分
corner casing　护角铸件
debris　碎片；残骸
efficiency curve　效率曲线
electrical generator　发电机
equivalent　等价的；相等的
extract　提取；取出；榨取
factory test　工厂测试
fan blade　风扇叶片；风机叶片
fitting　配件；装置
flat efficiency curve　平坦的效率曲线
flow inversion　流动逆转
gravitation　重力
grease　润滑油；油脂

hamster cage　仓鼠笼
illustration　说明；插图；例证；图解
impact　冲击；撞击
impinge　冲击；撞击
incoming water　来水；进水
installation　安装；装置
kW, kilowatt　千瓦
lab test　实验室测试
leakage　渗漏
lubricant　润滑油；润滑剂
lubrication　润滑
maintenance　维护
maximum efficiency　最大效率
mechanic　技工；机修工
modification　改进；变更；修改
nominal efficiency　名义效率
outer edge　外缘；外刃
overcome　克服；胜过
overlap　重叠；重复
partial load　部分负荷；分载
patent　*n*. 专利权；
　　　　v. 授予专利；取得……的专利权
peak efficiency　最高效率
performance　性能；绩效
power input　功率输入；电源输入
pressurized water　加压水
profiled steel　异形钢
profiled　异形
proportional　比例的；成比例的
purge valve　冲洗阀
regulating lever　调节杆；调整杆

regulating unit　调节装置
resemble　类似；像
resistance　阻力；电阻
resistant　有抵抗力的；抵抗的
rotating element　转动元件；转动部分
run-of-the-river plant　河床式水电站
schematic view　示意图
self-aligning roller bearing　调心滚子轴承
shaft power　轴功率
sharpen　削尖；磨快
shutdown　关机；停机
shutoff valve　截流阀；截止阀
silt　淤泥；泥沙
simplicity　简单；简易
slant　使倾斜；使倾向于
slide bearing　滑动轴承
split runner　拼合式转轮

spoon-shaped　勺形；匙形
squirrel cage turbine　鼠笼式水轮机
stainless steel　不锈钢
structural steel　结构钢；钢架
subdivided　再分，细分
suction head　吸水引高度；吸入水头
suction pressure　吸入压力
total efficiency　总效率
transversely　横着；横断地；横切地
trough-shaped　槽形的
turbine chamber　水轮机室
unattended　无人值守的
variation of speed　变速；改变转速
volume　量；体积
waste water　废水；污水
water wheel　水轮机；水车
yield　生产；屈服

4 Francis Turbines

4.1 Introduction

This turbine is named after J. B. Francis, an American engineer, who set about to improve the reaction type of turbines which existed in the year 1849.

The original Francis turbine has been modified from time to time to improve its efficiency, and, today, an efficiency of more than 93% has been reached.

This turbine is a versatile one and can work over a very wide range of heads, say, from 25 m onwards to 550 m.

This is a reaction type of turbine where water is admitted at high pressure into the runner, and pressure energy is converted into mechanical energy, because of the reaction of water when traveling over the curved vanes. As the pressure inside the turbine varies, this turbine is also called a variable pressure turbine. Another name for this turbine is the mixed flow turbine since water enters the runner radially, flows inwards, then changes the direction to the axial.

4.2 Main Components

The main components of the turbine are the scroll case or the spiral casing; the stay vanes; the guide vanes; the runner and the draft tube, as shown in Fig.4.1.

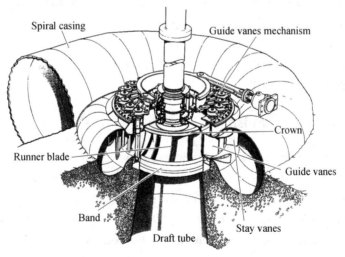

Fig.4.1 A Francis turbine

4.2.1 Scroll Case

This is an annular spiral-shaped casing forming the circumference of the turbine. One end of it is connected to the penstock to receive the water under pressure from it, and this water is admitted uniformly all over the circumference of the runner. The cross-sectional area of the casing decreases progressively, as more and more water is diverted on the runner so that the velocity of the flow is constant. The inner face of the casing is cylindrical, where the stay vanes ring is attached.

4.2.2 Stay Vanes Ring

This is a ring with a number of fin-shaped stay vanes welded between two annular rings, as shown in Fig.4.2. The inlet dimensions of the stay vanes are equal to the dimensions of the opening on the inner face of the scroll case. Stay vanes are also termed as fixed guide vanes.

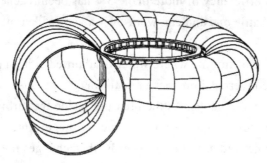

Fig.4.2 The scroll case with the stay vanes ring

The functions of the stay vanes are:
① To guide the water received from the scroll case at a proper angle on the movable guide vanes for which their shape is carefully designed.
② To strengthen the scroll case against the high pressure of the water passing through it. The stay vanes ring, therefore, is rigidly welded to the scroll case. The scroll case and stay vanes are surrounded by concrete after they are erected.
③ To transmit the load of the main thrust bearing of the generator to the foundations.
④ To provide connections for the head cover.

4.2.3 Guide Vanes Mechanism

The guide vanes mechanism (Fig.4.3) serves three functions:
① To receive water from the stay vanes and direct it at proper angles on the runner vanes so that the shock at the entry of the runner is minimum.
② To govern the turbine by controlling the quantity of water admitted to suit the instantaneous load conditions. This is achieved by rotating the guide vanes about their individual axis and varying the cross sectional area of the openings between the two neighboring vanes.
③ To serve as a valve, stoping the flow of water in a closed position.

The guide vanes are located between the stay vanes and the runner. The vanes are aerofoil-shaped so that water flows around them without separation. In the closed position, the toe of any guide vane must fit the heel of the next vane exactly so that there is no leakage. Each guide vane is provided with a shaft about which it rotates. The lower position of this shaft rests in a small bearing fitted in a housing which is attached to the bottom ring. The upper portion projects the head cover through a guide bearing. All these protruding shafts are connected by means of links and levers to an annular ring called the distributor ring. A slight clockwise or anticlockwise movement of the ring rotates all the guide vanes identically, thereby, opening or closing them. The distributor ring is moved by two piston servomotors attached tangentially 180' apart to the ring. The governor mechanism moves one piston forward and other backward to an equal extent (similar to the movements of hands guiding the driving wheel of a car).

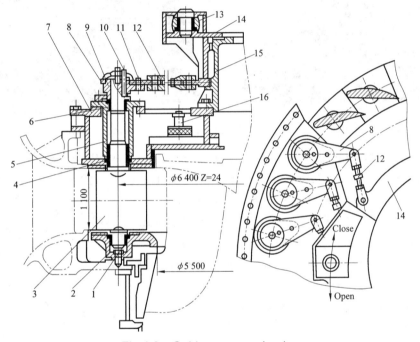

Fig.4.3 Guide vane mechanism

1, 4, 6—nylon bearing bush; 2—bottom ring; 3—guide vane; 5—axle sleeve; 7—head cover; 8—connection plate; 9—steering arm; 10—split key; 11—shear pin; 12—connecting rod; 13—push-and-pull rod; 14—distributor ring; 15—support; 16—air compensating valve.

The number of moveable guide vanes can be much more than that of the runner vanes. Normally between 18 and 26 vanes are provided but in some cases the number can be as many as 32.

4.2.4 Runner

This is the heart of the turbine. The runner is located between the guide vanes and the draft tube. Water enters the runner over the entire periphery, imparts most of its energy while traveling over the runner vanes and then enters the draft tube.

The direction of the flow of water is radially inwards at the entry to the runner and gradually changes to the axial direction by the time it reaches the exit of the runner. The runner consists of three main parts: the crown, the vanes and the band, as shown in Fig.4.4. The shape of the runner vanes has to be carefully designed for maximum conversion of energy. Hydraulic friction and eddy current losses have to be kept to a minimum. The upper edges of the vanes are attached to the crown at the extreme streamlines of the flow. On the outer face of the crown, holes are provided for fitting it to the runner shaft. The band surface also conforms to a streamlined pattern if the inlet diameter of the runner is larger than the outlet diameter. If it is smaller, then the band is annular in shape.

Fig.4.4　Runners of a low head (left) and a high head (right) Francis turbines

The shape of the runner depends on the head, output and specific speed of a turbine.

The number of vanes of a runner depends upon head, output and rotational speed of the runner. It may vary from 10 to 20. This number may be less than the number of guide vanes.

To prevent water under pressure from bypassing the runners and finding its way directly from the guide vane to the draft tube, sealing rings called labyrinth rings are provided at the exit of the runner on the outer face of the band, as shown in Fig.4.5.

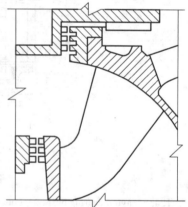

4.2.5　Draft Tube

Fig.4.5　Labyrinth rings

It is located at the outer of the runner. One end of it is at the runner outlet and the other end in the tail race. The draft tube serves the following functions:

① It carries the water, discharging from the runner, to the tail race.

② It creates a negative pressure at the runner outlet, increasing the effective head on the runner. The total head acting on the runner blades is thus the sum of the static head (equal to the difference between the headwater level and the elevation of the runner exit section), and the negative head (vacuum) beneath the runner. This helps locate the runner above the tail race obviating the need for costly foundation work.

③ It recovers the velocity energy by converting it into pressure energy, thereby increasing the efficiency of the turbine. The velocity head at the runner outlet can be 10% to 30% of the effective head. The shape of the draft tube must be carefully designed for optimum condition.

The most common shape is an elbow shape with a conical section followed by a bend and then the diverging section with an oval or rectangular cross-section, as shown in Fig.4.6.

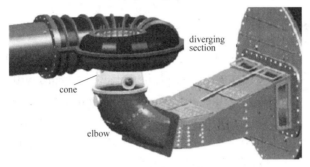

Fig.4.6 An elbow shape draft tube

The ratio of outlet area to inlet area is large if the recovery has to be large. If the outlet area is very large, then it is divided into a number of portions with dividing walls along the flow direction.

For turbines working under low heads, the outlet diameter is larger than the inlet diameter. The runner can be removed from the downward direction for inspection and repairs. A detachable section of the draft tube has to be provided.

After the draft tube liner is erected, it is covered with concrete and equipped with necessary fasteners, ribs and so on. Often the exit section of the draft tube can be made of concrete itself. An inspection gate is provided to enter the draft tube whenever needed.

4.2.6 Head Cover

This is a cylindrical box-like structure provided on the top surface of the runner, as shown in Fig.4.7.

Fig.4.7 Head cover of a Francis turbine

Its functions are ① to prevent any water leaking through the gap between the guide vanes and the runner from finding its way outside the turbine, ② to provide bolts on the periphery for connecting it to the main turbine body, ③ to provide circular openings for fitting the guide bearings for the guide vanes trunnions projecting out, ④ to provide an opening in the center for fitting the main guide bearing of the turbine shaft and the stuffing box, the diameter of the opening must be bigger than the diameter of the flange of the shaft, and ⑤ to provide mounting of the distributor ring and other accessories.

4.2.7 Bottom Ring

This is located at the bottom of the turbine guide vanes between the stay vane ring and the draft tube, as shown in Fig.4.8. It has an upper annular ring with an external diameter equal to the internal diameter of the stay vanes ring and a lower annular portion with a diameter approximately equal to the diameter of the draft tube. Holes are provided in the annular ring for the lower end of the guide vanes trunnion to pass to the bearings.

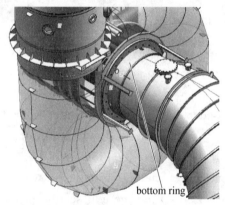

Fig.4.8　Bottom ring of a Francis turbine

4.2.8　Shaft

The shaft disposition of most of the modern Francis turbines is vertical so that the generator can be kept above the highest tail water level. The danger due to flooding can thus be avoided and the turbine can be kept at another level decided by its design parameters. In the powerhouse, there can be different floors for generators and turbines so that erection work can go on independently.

The turbine shaft can be directly bolted to the generator shaft. If the shaft length is long, an intermediate shaft is provided so that the turbine runner can be removed by dismantling this intermediate shaft, without disturbing the generator shaft.

There is only one thrust-cum-guide bearing provided for the entire hydraulic unit consisting of the turbine and the generator in case of a vertical layout. This bearing is generally below the generator rotor or, in some cases, above the casing of the turbine. There are two guide bearings provided—one at the top of the generator shaft and the other at the top of the turbine shaft.

The horizontal shaft layout is presently adopted for the Francis turbine forming part of the hydro units used for pumped storage plants, and each unit consists of a turbine, a generator motor and a pump. This layout is rarely used for pure power generation. The runner is fitted to the main shaft in an overhung position. If the bearings are provided to the runner, the shaft is required to pass through the draft tube. This is found to be an advantage especially during partial load operations as it avoids creating a vacuum at the bend.

4.2.9 Turbine Pit Liner

This is a cylindrical liner with an adequate diameter to permit the removal of the turbine head cover as a single piece and sufficient height to extend up to the turbine floor from the top of stay vanes ring, as shown in Fig.4.9. Concreting is done on the outside of the liner and the liner acts as a framework. Necessary ribs, bolts and other fasteners are provided on the outer surface of the liner.

Fig.4.9 Already mounted pit liner on the spiral case

4.2.10 Dewatering of Turbine

To dewater a small turbine, a valve is provided on the upstream face and draft tube gates are provided at the end of the draft tube. By closing these two components, the turbine space can be dewatered. For a large turbine, a valve cannot be provided. So a gate is provided just after the entrance of the penstock pipe to control the flow of water.

Questions

1. Given a drawing of a Francis turbine, identify the following major components:
   · Spiral casing
   · Guide vanes
   · Stay vanes
   · Runner
   · Draft tube
2. Given a drawing of a Francis turbine runner, identify the following major components:
   · Crown

· Vanes

· Band

3. State the purposes of the following Francis turbine components:
 · Spiral casing
 · Guide vanes mechanism
 · Stay vanes ring
 · Runner
 · Draft tube
 · Crown
 · Vanes
 · Band

4. Given a drawing of different Francis turbine runners, identify the one with higher specific speed, the one with higher head and lower flow rate.

References

[1] Arne Kjølle. Hydropower in Norway-Mechanical Equipment [M]. Trondheim: Norwegian University of Science and Technology, 2001.

[2] Magdy Abou Rayan, Nabil H. Mostafa, Purage Ohans. Textbook of Machines Hydraulic [M]. Zagazig: Zagazig University, Dept. of Mechanical Engineering, 2009.

[3] Ingram G. Basic concepts in turbomachinery[M]. Frederiksberg: Grant Ingram & Ventus Publishing ApS, 2009.

[4] GE Renewable Energy. Francis Hydro Turbine [EB/OL]. [2016-06-16]. https://www.gerenewableenergy.com/hydro-power/large-hydropower-solutions/hydro-turbines/francis-turbine.html.

Important Words and Phrases

accessory　配件；附件
admit　进入
aerofoil shaped　翼形；机翼的形状
air compensating valve　补气阀
annular ring　环孔；环状垫圈
anticlockwise　逆时针
axle sleeve　轴套
bearing bush　轴瓦
bolt　螺栓
bolt　用螺栓固定
bottom ring　底环
bypass　绕过；绕开
clockwise　顺时针

conform　符合；遵照
conical section　锥管段
connecting rod　连杆
connection plate　连接板
cross-sectional area　横截面面积
crown　上冠
design parameter　设计参数；设计规范
detachable　可拆式；可分开的；可拆开的
dewater　排水
distributor ring　控制环
diverging section　扩散段
divert　导向；使转移
dividing wall　隔墙；分隔墙

driving wheel 驱动轮；主动轮	outlet diameter 出口直径
eddy current 涡流	oval 椭圆形
effective head 有效水头	overhung 悬臂式的
elbow shape 肘形	part load 部分负荷
erect 安装；建造	piston servomotor 活塞接力器
external diameter 外径	powerhouse 厂房
fastener 紧固件；扣件	progressively 渐进地；逐步地
fin-shaped 鳍状；鳍形	protrude 使突出；使伸出
fixed guide vane 固定导叶	pumped storage plant 抽水蓄能电站
framework 框架；骨架；结构	push-and-pull rod 推拉杆
gap 间隙；缺口	recover 回收；恢复；弥补
generator rotor 发电机转子	rib 肋拱；肋材
governor mechanism 调速机构；调节器	rotational speed 转速
guide bearing 导轴承	runner vane 转轮叶片；轮叶
guide vanes mechanism 导水机构	sealing ring 密封环；密封圈；垫圈
head cover 顶盖	set about 着手；开始做……
headwater 上游水位；上游源头	shear pin 剪断销
identically 同一地；相等地	spiral casing 蜗壳
improve 改进；改善；改良；增进	spiral-shaped 螺旋状
inlet area 进口面积	split key 分半键
inlet diameter 进口直径	static head 静压头；落差
inner face 内表面	stay vane 固定导叶
inspection gate 检查口；检测门	stay vanes ring 固定导叶环
intermediate 中间的	steering arm 转臂
internal diameter 内径	streamline 流线；流线型
labyrinth ring 迷宫环；曲折密封圈	strengthen 加强；巩固
leakage 渗漏；渗漏物；漏出量	stuffing box 填料盒；填料函
liner 衬垫；衬套；衬里	tail water level 尾水位
links and levers 连杆和转臂	thrust bearing 推力轴承；止推轴承
movable guide vane 活动导叶	thrust-cum-guide bearing 推导组合轴承
negative head 负水头；负压头	total head 总水头；总压头
negative pressure 负压	trunnion 耳轴
nylon 尼龙	upstream face 上游面
obviate 排除；避免；消除	vacuum 真空
onward 向前的	variable pressure turbine 变压力水轮机
opening 开口	velocity energy 速度能
outer face 外面	versatile 多用途的；多功能的
outlet area 出口面积	

5 Propeller and Kaplan Turbines

The propeller and Kaplan turbines are reaction turbines of the axial flow type. The difference between the propeller and Kaplan turbines is that the propeller turbine has fixed runner blades while the Kaplan turbine has adjustable runner blades.

5.1 Development

5.1.1 Development of Propeller Turbines

At many power plant sites, where large quantities of water were available at low heads, the dimensions of Francis turbine started becoming too large and uneconomical, and another turbine named propeller turbine was developed.

This turbine has major components like the scroll case, stay vanes, guide vanes and the draft tube similar to the Francis turbine, as shown in Fig.5.1. However, the runner differs in the following respects:

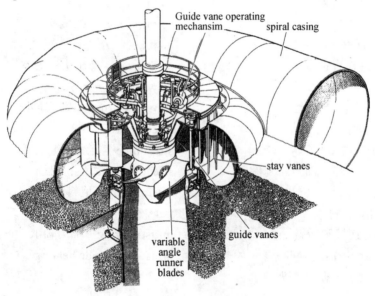

Fig.5.1 Kaplan turbine

① The runner is of the axial flow type and water enters and leaves the runner blades parallel to the axis while the Francis turbines runner is of the mixed flow type.

② The runner vanes resemble the blades of a propeller. They are cantilevers mounted on the runner hub and the outer edges are not connected to each other (the runner does not have lower band).

③ The number of blades varies between 4 and 8, considerably less than the Francis turbine, therefore, a lesser area is occupied by the blades and more area is available for flow.

The propeller turbine was found to have a very high efficiency at or near the rated output but its efficiency at partial loads was low. Consequently, the use of this turbine had to be restricted to plants with constant output.

5.1.2 Development of Kaplan Turbines

Many attempts were made to improve the partial load efficiency of propeller turbines and Kaplan developed a successful design where the runner blades were also made movable and their angle of inclination could be changed to suit the instantaneous load conditions. The Kaplan turbine was named after Professor Victor Kaplan (1876-1924). The turbines work in a head range of 5 m to 60 m and in exceptional cases up to 80 m. The specific speed ranges from 300 to 1,000.

As the inclination of the runner blade has to be changed constantly, it becomes imperative to provide a mechanism similar to the guide vanes operating mechanism. As the runner vanes are rotating all the time, the governing mechanism must also rotate all the time.

5.2 Main Components of the Runner

The main components of the runner are: ① the hub; ② the vanes with operating mechanism; and ③ the nose cone (Fig.5.2).

5.2.1 Hub

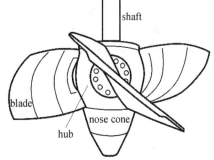

Fig.5.2   The runner of a Kaplan turbine

The hub is a hollow cylindrical piece with spherical midriff. The runner vanes are mounted on the specially provided recesses on the periphery of the hub; with their trunnions projecting into the body of the hub. The operating mechanism is provided inside the body of hub. The hub is completely filled with oil to provide reliable lubrication of moving parts. The oil pressure inside the hub is kept higher than the outside water pressure to prevent water penetration into the oil.

The nose cone is fitted on the end of the hub at the draft tube side. This cone can be in a single piece or in two pieces depending upon the design. The nose cone is designed to offer minimum resistance to the flow. Necessary fasteners are provided on the periphery of the hub for connecting the cone. The other end of the hub is connected to the turbine shaft.

5.2.2 Blades

The blade consists of the following parts: a wing-shaped cantilever portion forming an arc of an annular ring, which receives the water and a trunnion which supports the wing-shaped portion and transmits the hydraulic thrust and any centrifugal force developed to the hub.

The blades have to be sufficiently wide so that even in a fully open position there is sufficient overlap with the neighboring blades and no water is allowed to pass directly to the draft tube through the gaps between two blades.

The number of blades, and thus the diameter of the runner hub, increases with the head. Four blades are ordinarily used at heads of up to 20 m, while eight blades are used at heads of between 40 and 80 m.

The blades can be bolted to the trunnions attached to the levers forming a part of the controlling mechanism. Special packings are provided at the hub recesses to avoid any leakage of water inside the hub and any leakage of lubricating oil filled inside the hub at a pressure somewhat higher than that of the maximum head. If, by chance, water finds its way into the runner hub, it will settle down at the bottom of the hub and will be expelled by the air pressure through specially provided outlets.

The inclination of the runner vanes can be varied from an angle of 0° to 30° by means of a runner blade operating mechanism–the operating time required being 15 to 30 seconds.

5.2.3 Runner Blade–Operating Mechanism

Its function is to vary the runner blade angle to suit the instantaneous load conditions in coordination with the guide vane-operating mechanism. It comprises the following parts: servomotor, main blade-operating spindle, cross head, toggle levers, trunnion, and pressure oil head, as shown in Fig.5.3.

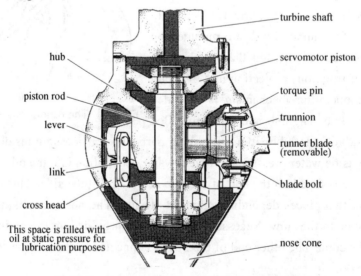

Fig.5.3 A runner blade operating mechanism

The servomotor is a cylinder inside which the servomotor piston is moved axially by means of oil pressure. The blade-operating rod, rigidly connected to the piston, moves the toggle levers and links to vary the angle of the runner vanes. Kaplan turbine is equipped with an oil head on the top of the generator or exciter and with vertical, rotating oil pipes housed within the shaft of the generator and turbine. The oil head conveys the pressure oil from the stationary governor oil pipes to the rotating pipes which lead to the runner blade servomotor.

An example of the regulating system of the runner blade slope is shown in Fig.5.4. The slope of the runner blades (1) are adjusted by the rotary motion activated by the force from the piston (4a) through the rod (5). The cylindrical extension of the upper end of the turbine shaft (6) serves as a servomotor cylinder (4) whereas the lower flange of the generator shaft (8) serves as a cover. The rod (5) moves in the two bearings (7).

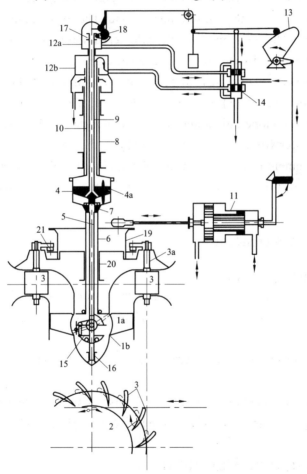

Fig.5.4 Regulating mechanism of guide vanes and runner blades

1—runner vanes; 2—guide vane cascade; 3—guide vanes; 4—runner vane servomotor; 4a—servomotor piston; 5—rod for force transfer from 4a; 6—turbine shaft; 7—bearings; 8—generator shaft; 9 and 10—coaxial pipes; 11—servomotor for guide vane cascade; 12—oil inlet; 12a and 12b. oil chambers; 13—cam; 14—regulation valve for high pressure oil to runner servomotor;15—lever spider for force transfer from 5; 16—conveying of rod 5;17—lining for feedback lever; 18—feedback lever; 19—regulating ring; 20—radial bearing; 21—guide vane lever

The oil supplied to the servomotor (4) is entered at the upper end of the generator shaft (8). The oil is conveyed to the respective sides of the servomotor through two coaxial pipes (9) and (10) inside the hollow generator shaft. The inner tube (9) conveys oil to and from the lower side of the piston (4a), whereas the annular opening between the pipes (9) and (10) conveys oil to and from the piston top side. The oil is supplied through the entrance arrangement (12) with the two chambers (12a) and (12b) at the top of the unit.

5.2.4 Cooperation of Guide Vanes and Runner Blades Regulation

There is a correlation between the runner vane angle and guide vane angle to obtain the best efficiency conditions over the entire range of load. The exact nature of the relationship is a function of the head. The movements of the two servomotors are automatically adjusted to optimum for all head and load variations by means of a governing device.

The turbine governor operates directly on servomotor (11) which executes the movement of the guide vanes (3) as shown in Fig.5.4. The movement of the servomotor triggers and controls the slope adjustment of the runner vanes. This is carried out by a rod and lever which transfer from the servomotor (11) to the cam (13), and the cam is turned according to the movement of the servomotor piston (11). In this way the spool valve (14) is moved out of the neutral position and the servomotor piston (4a) is then moved by the oil pressure supply.

5.2.5 Location of Servomotor

There are three possible locations:

① In the hub of the runner as in the case of the Lasalle power station, Sweden. Though this location gives a shorter piston rod, the size of the servomotor becomes unwieldy at higher heads.

② Between the turbine and the generator shafts. This method increases the dimension of the unit.

③ At the top of the generator shaft as in the case of the Tres Marias plant, Brazil. This location gives a compact hydro-unit, though it needs a long piston rod.

5.3 Scroll Case and Stay Ring

Water from the penstock enters the scroll casing surrounding the runner, as shown in Fig.5.5. The scroll case provides an even distribution of water around the circumference of the turbine runner maintaining the constant velocity of the distributed water.

For a turbine having heads less than 30 m, the scroll case is made of concrete with a trapezoidal cross-section.

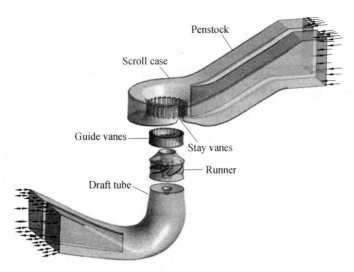

Fig.5.5　General arrangement of a Kaplan turbine

For higher heads, the hydraulic pressure may be too high for the concrete to withstand the load. In such cases, scroll casings of steel plates are designed in a way analogous to that of Francis turbines. The cross sections of the scroll casing are normally of circular shape and the steel plate shells are welded to the stay ring.

The stay ring is provided even when the scroll case is made of concrete. The vanes in the stay ring direct the water from the scroll casing to the guide vanes. In addition, the hydraulic forces are transferred through the stay ring and the stay vanes which are anchored to the concrete with large prestressed stay bolts. The stay vanes are normally made of welded steel plates and filled with concrete.

5.4　Guide Vane Cascade

The guide vane cascade of Kaplan turbines are constructed in the same way as Francis turbines except that the height of the vanes is more and the c to c distance between the opposite vanes is less. In the sense of operation, a regulating ring rotates the guide vanes through the same angles simultaneously when adjustments follow changes of the turbine load. The vanes are made of steel plate material and the trunnions are welded to them. The vane is designed purposely to obtain optimal hydraulic flow conditions, and they are given a smooth surface finish.

5.5　Head Cover and Draft Tube

As the physical dimensions of the turbine increase, the head cover is divided into three. The outer portion has holes for the projecting out of guide vane trunnions. The intermediate portion has the shape of a hollow hyperboloid and the inner portion is combined with the runner hub by a flanged connection, as shown in Fig.5.6.

A draft tube is a pipe with a gradually increasing area used for discharging water from exit of turbine to the tail race. By using the draft tube, the kinetic energy rejected at the outlet of turbine is converted to useful pressure energy.

The draft tube of a Kaplan turbine is proportionately larger. The velocity head at the exit of the runner may be many times as large as 40% of the effective head. As the draft tube is required to recover most of the head, its design assumes great significance. The area of cross-section at the exit is very large and it may be divided into two or three segments. It is elbow-shaped and can be made entirely in concrete with steel plate liners at the inlet section.

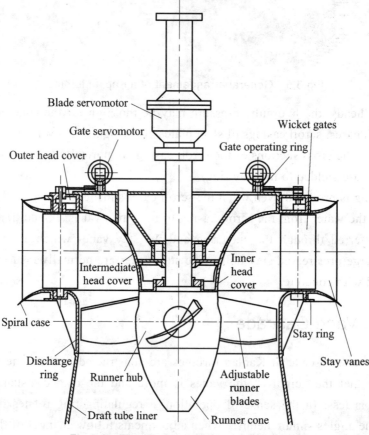

Fig.5.6 The Kaplan turbine configuration

5.6 Automatic Air Valve

It is provided on the top cover to destroy any vacuum that may be created above the top of the runner due to the sudden closure of guide vanes on load rejection.

5.7 Shaft of the Hydro Unit

This is hollow to accommodate the runner vane-operating mechanism.

5.8 Over-speed Protective Device

If the speed of the turbine exceeds the synchronous speed by more than 30%, the overspeed switch shuts down the unit and simultaneously opens the runner vanes fully to reduce the overspeed.

Questions

1. List the differences between Francis turbine runners and Kaplan turbine runners.
2. State the purposes of the following Kaplan turbine components:
- Guide vane cascade
- Stay ring
- Scroll case
- Head cover
- Runner hub
- Nose cone
- Runner blade
- Draft tube
3. Given a drawing of a Kaplan runner, identify the following major components:
- Shaft
- Hub
- Nose cone
- Servomotor
- Piston rod
- Blade
- Trunnion
- Link and lever
4. Explain the regulating process of the runner blade slope.
5. Explain the combined regulating process of guide vanes and runner blades.
6. List the possible locations of servomotor, and give examples.

References

[1] Arne Kjølle. Hydropower in Norway-Mechanical Equipment [M]. Trondheim: Norwegian University of Science and Technology, 2001.

[2] Magdy Abou Rayan, Nabil H. Mostafa, Purage Ohans. Textbook of Machines Hydraulic [M]. Zagazig: Zagazig University, Dept. of Mechanical Engineering, 2009.

[3] Ingram G. Basic concepts in turbomachinery[M]. Frederiksberg: Grant Ingram & Ventus Publishing ApS, 2009.

[4] GE Renewable Energy. Kaplan Hydro Turbine [EB/OL]. [2016-06-16]. https://renewables.gepower.com/hydro-power/large-hydropower-solutions/hydro-turbin

es/kaplan-turbine.html.

[5] R.E.A. Arndt, Hydraulic turbines, in the Engineering Handbook[M]. 2nd ed., Oxford: CRC Press LLC, 2005.

[6] Gagnon J M, Aeschlimann V, Houde S, et al. Experimental investigation of draft tube inlet velocity field of a propeller turbine[J]. Journal of Fluids Engineering, 2012, 134(10): 101102.

Important Words and Phrases

adjust 调整；使……适合
air pressure 气压；风压
analogous 类似的
anchor 使固定
angle of inclination 倾角；倾斜角
arc 圆弧
assume 承担；假定；采取；呈现
attach 把……固定；贴上
automatic air valve 自动空气阀；自动排气阀
cam 凸轮
cantilever 悬臂
centrifugal force 离心力
closure 关闭；终止
coaxial 同轴的；共轴的
correlation 相关；关联；相互关系
cross head 十字头；丁字头
elbow-shaped 肘形
exceptional 异常的；例外的
exciter 励磁机
execute 实行；执行
expel 排出；驱逐；开除
feedback lever 反馈杆；反馈杠杆
flange 法兰
function 函数
governing device 调节装置
governing mechanism 调节机构；调速机构
guide vane cascade 导叶叶栅
guide vanes operating mechanism 导叶操作机构
hollow 空的；中空的

hydraulic thrust 水推力；液压推力
hyperboloid 双曲面；双曲线体
imperative 必要的；不可避免的
in coordination with 配合；与……协调
inlet section 进口段；进口部分
lever 转臂；操作杆；杠杆
load rejection 甩负荷
lubricating oil 润滑油
mechanism 机制；机能；机理；机械装置
midriff 中腹部
movable 可移动的；不固定的
nose cone 前锥体；泄水锥
oil head 受油器
oil pipe 油管
operating mechanism 操作机构
operating spindle 操作杆
operating time 操作时间；作业时间
over-speed protective devices 超速保护装置
overspeed switch 超速开关
packing 填料；包装；填充物
penetration 渗透；突破；侵入
piston rod 活塞杆
plant 电站
power plant 发电站；电厂
prestressed 预应力
project into 深入；插入
propeller turbine 定桨式水轮机
rated output 额定输出；额定功率
regulating ring 控制环
regulation valve 调节阀
rigidly 刚性的；不易弯曲的

scroll casing　蜗壳
segment　部分
servomotor piston　接力器活塞
site　现场；地点；位置；场所
slope　斜率；倾斜；斜坡
spider　支架；三脚架
spool valve　柱形阀；短管阀
stay bolt　拉杆螺栓
stay ring　座环

synchronous speed　同步转速
toggle lever　曲柄连杆
torque pin　扭力销；转矩销
trapezoidal　梯形的
trigger　引发；引起；触发
uneconomical　不经济的；浪费的
unwieldy　笨拙的；笨重的
wing-shaped　翼状的

6 Deriaz Turbines

6.1 Introduction

Deriaz Turbine is also known as diagonal turbine. This turbine was named after its inventor Paul Deriaz, a hydraulic designer. It is basically a modified Kaplan turbine, adjusted for a large number of rotor blades. This is a turbine with water passages inclined to the shaft with movable vanes runner, as shown in Fig.6.1. The Deriaz turbine applies for heads up to 200 m, with an efficiency in excess of 92%. It can also be adopted as a reversible pump turbine with a pumping head of 60 m. The Deriaz design has been proved useful for higher heads and also for some pumped storage applications.

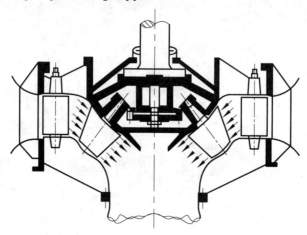

Fig.6.1 Schematic view of a Deriaz turbine

The Deriaz turbine is classified as a double-regulated, reaction, mixed-flow turbine. This innovative product has been built by Geppert for small hydro power plants since 1999 which has been successfully installed in dozens. The first non-reversible Deriaz turbine, capable of producing 22,750 kW with a head of 55 m, was installed in an underground station at Culligran, Scot., in 1958. An early Deriaz turbine system was installed at plants on both the Canadian and U.S. sides of Niagara Falls.

Adjustable runner and guide vanes maintain a high efficiency over a wide flow range for this medium pressure turbine with resulting high partial-load efficiencies, comparable with Kaplan turbines. The efficiency of a diagonal flow turbine is slightly less than the peak efficiency of a Francis turbine, but a Francis turbine only has a high efficiency within the design range of heads and loads. On the contrary, the blades of a diagonal flow turbine can be adjusted to suit the head and load changes, so the overall efficiency is higher. With seasonal strongly variable discharge, a higher annual production can be reached with this turbine type than with comparable Francis turbines.

This turbine type is often installed with seasonal strongly variable discharge in order to avoid costly solutions with more Francis units. Also high positioned Kaplan turbines have been replaced by diagonal turbines as these have turned out to be less cavitation sensitive.

6.2 Development

The trend towards higher-head Kaplan turbines in excess of 60 m head was obvious as the hydropower potential was increased after World War II. If the head on a Kaplan turbine is increased, the following complications arise:

① The hub diameter increases and the radial dimensions of the runner blades decrease.

② The number of blades increases and so do the links and levers.

The hollow space inside the hub is not adequate to accommodate the bigger blade-operating mechanism. With the conventional, axially arranged Kaplan turbine the runner hub would have become so large and the vanes so narrow that an appreciable loss in turbine efficiency resulted. The logical step was to revert to the mixed flow arrangement of the Francis turbine, resulting in a Francis turbine with movable blades, as shown in Fig.6.2.

Fig.6.2   A Deriaz turbine runner

The scroll case, stay vanes, guides vanes and the draft tube were kept the same as in the Kaplan turbine but the water passages from the guide vanes to the draft tube were made inclined to the shaft. Instead of being at right angles to the shaft, the axes of the runner vanes were inclined to the shaft and perpendicular to the water passage, so that the hydraulic limitations of the Kaplan turbine for high heads are overcome. The direction of flow was diagonal as against axial in the case of Kaplan turbine.

The inclined blades make the Deriaz turbine particularly suitable for the head range between 20 m and 100 m—a window of operation between that of the Kaplan and Francis turbines. Because the runner blades are adjustable, the Deriaz design offers a number of other advantages: smooth and efficient operation over a wide range of head and load, uniform distribution of pressure and load across the blade (i.e., from the casing to the mid-span to the hub), freedom from development of cavitation across the entire operating range, lower specific and runaway speed, and lower hydraulic thrust.

6.3 Mechanical Considerations

In an axial flow turbine, the runner hub is cylindrical, and the axes of blades are perpendicular to the runner shaft. A distinguishing feature of a diagonal turbine is the fact that the axes of the vanes lie at an acute angle to the turbine's shaft of rotation, and the conical rotor hub does not constrict the flow, thus making it possible to increase the number of vanes and to use these turbines at higher pressures. Here, the flow of water against the turbine blades is neither axial nor radial, but at an angle. The angle θ made by the blade axes with the main shaft decreases with increasing head and ranges between 30° and 60°, as shown in Fig.6.3.

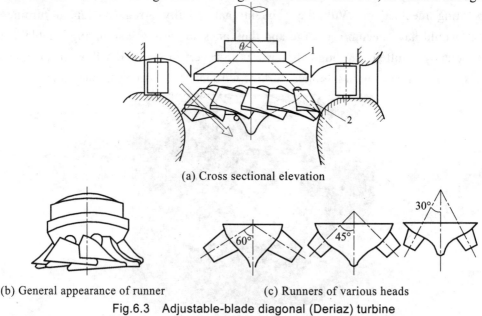

(a) Cross sectional elevation

(b) General appearance of runner   (c) Runners of various heads

Fig.6.3　Adjustable-blade diagonal (Deriaz) turbine

1—runner hub; 2—blades

The runner hub is conical in shape and it is much shorter in height compared to the Kaplan turbine. But its base is much larger and there is adequate space for the lever and link mechanism for increased number of blades.

The servomotor design can be considerably simplified.

6.4 Hydraulic Considerations

Because of the larger inlet area, the velocity is low. As the inlet linear velocity is more than outlet linear velocity, it is possible to make the inlet and outlet runner angles the same. This improves the partial load efficiency and reduces the tendency towards cavitations.

In a Kaplan turbine, the water has to make a 90° turn before entering the runner chamber, which will affect the flow velocity and reduce the water energy. Unlike a Kaplan turbine, the inclined water passage of a Deriaz turbine makes the flow pattern more direct and there are fewer pronounced changes in the flow pattern. The surface area of the runner is less and both these factors lead to reduction in frictional losses.

Comparative characteristics of a diagonal hydraulic turbine and a radial-axial hydraulic turbine are presented in Fig.6.4, where η/η_{max} is the ratio of the efficiency in operating modes to the maximum efficiency; N/N_{opt} is the ratio of the power in operating modes to the optimal power. Because of the better flow around the vanes of the hub and the suction pipe in modes that differ noticeably in load and head from calculated values, flow conditions are calmer in a diagonal hydraulic turbine, have smaller pulsations, the efficiency curve is gentler, and the average operating efficiency η is higher. The cavitational characteristics of a diagonal hydraulic turbine are somewhat worse than that of radial-axial turbines. Therefore, diagonal turbines can be installed at hydroelectric power plants with heads of up to 200 m, displacing radial-axial hydraulic turbines in this range. Diagonal hydraulic turbines are especially economical at plants that have large fluctuations in pressure head and power.

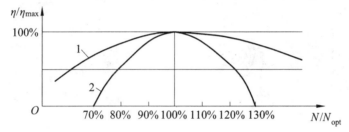

Fig.6.4 Comparative characteristics of diagonal (1) and radial-axial (2) turbines

6.5 Runaway Conditions

The Deriaz turbine has reduced discharge at overspeed. If the turbine starts racing to runaway speed, the centrifugal head built decreases the flow, and this tendency increases with an increase in the speed of the turbine. This self-governing tendency leads to a lower runaway speed (sudden loss of load) than a Kaplan turbine and, consequently, the generator can be made smaller, which results in significant savings in generator costs. This also reduces the size of the overhead crane and there is sufficient saving in the cost.

6.6 Reversibility

The rotors of diagonal turbines are also used extensively in the manufacture of reversible hydraulic machinery (pump turbines) for pumped-storage electric power plants. The Deriaz turbine can work as a pump for heads up to 90 m whereas the Kaplan turbine can pump against a maximum head not exceeding 20 m. Its first installation was in the Sir Adam Beck 2 pumped storage power station of the hydroelectric power commission of Ontario, Canada, where six units were installed in the late 1950s.

6.7 Servomotor

The design of the runner blade operating mechanism is quite cumbersome in Kaplan turbines. For high head turbines, it has to be placed outside the runner, either between the turbine and the generator shaft or on top of the generator. This needs a long and heavy piston rod.

In the Deriaz turbine, the mechanism for adjusting the blade angles is located within the hub, and a simpler servomotor mechanism of rotating shaft type can be provided, as shown in Fig.6.5.

 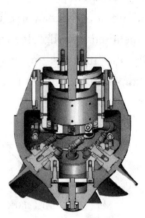

Fig.6.5 The runner and eight inclined blades regulation system of a Deriaz turbine

Rotating movement of the servomotor piston is transmitted to the spider inside the runner hub through the piston rod. Slide block is provided for each connection of the spider and the blade arm, and it conveys the rotating motion of the spider to the runner blade arm. Thus the angle of attack of the blade changes to suit the head and load variations.

Questions

1. List the differences between Deriaz turbines and Kaplan turbines.
2. List the advantages of Deriaz turbines.
3. Explain the improvement in Deriaz turbines in reducing the friction losses.

References

[1] Diagonal Hydroturbine. (n.d.) The Great Soviet Encyclopedia, 3rd Edition. (1970-1979)[EB/OL]. [2016-06-16]. http://encyclopedia2.thefreedictionary.com/Diagonal+Hydroturbine.

[2] 高建铭, 林洪义, 杨永骅. 水轮机及叶片泵结构[M]. 北京: 清华大学出版社. 1992.

[3] Academic. Hydraulicians in Europe 1800-2000 DÉRIAZ [EB/OL]. (2013) [2016-06-16]. http://hydraulicians.enacademic.com/703/D%C3%89RIAZ.

[4] Ingram G. Basic concepts in turbomachinery[M]. Frederiksberg: Grant Ingram & Ventus Publishing ApS, 2009.

[5] OMOS Machine Tools. Turbine Models [EB/OL]. [2016-06-16]. http://www.omos.cz/_en/turbine-models/.

Important Words and Phrases

acute angle　锐角
angle of attack　攻角
annual production　年产量
appreciable　相当可观的
base　基座；基础；底部
blade arm　叶片臂
blade-operating mechanism　叶片操作机构
cavitational characteristic　空化特性
civil work　土建工程；土建工作
complication　并发症；复杂；复杂化
diagonal　斜的；对角线的
double-regulated　双重调节
electrical equipment　电气设备；电力设备
fluctuation　起伏；波动
frictional loss　摩擦损失
innovative　创新的
linear velocity　线速度
mid-span　中跨
movable vane　可动叶片
non-reversible　不可逆的
overall efficiency　总体效率
overhead crane　桥式吊车；高架起重机
pulsation　脉动
pump turbine　水泵水轮机
pumped storage　抽水蓄能
pumping head　抽送扬程
racing　空转
reversibility　可逆性
reversible hydraulic machinery　可逆式水力机械
reversible pump turbine　可逆式水泵水轮机
revert　恢复；使恢复原状；重提
right angle　直角
rotating shaft　转轴；旋转轴；回转轴
runaway condition　失控工况；飞逸工况
runaway speed　飞逸转速
slide block　滑块
water passage　流道

7 Bulb Turbines

7.1 Introduction

The bulb turbine is a reaction turbine of Kaplan type which is used for the lowest heads. It is characterized by having the essential turbine components as well as the generator inside a bulb, from which the name is developed. A main difference from the Kaplan turbine is that the water flows with a mixed axial-radial direction into the guide vane cascade but not through a scroll casing. The guide vane spindles are inclined (normally 60°) in relation to the turbine shaft. Contrary to other turbine types, bulb turbines are shaped in a conical guide vane cascade.

The bulb turbine runner is of the same design as for the Kaplan turbine, and it may also have different numbers of blades depending on the head and water flow.

7.2 General Arrangement

A general arrangement of a bulb turbine plant is shown in Fig.7.1 by a vertical section through the unit. Fig.7.2 shows the turbine design in more detail.

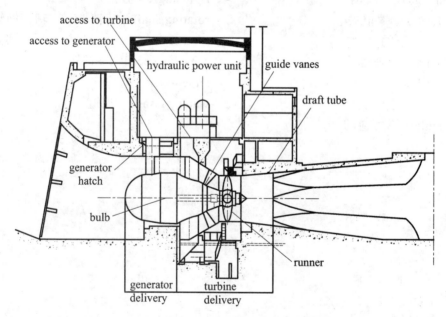

Fig.7.1 General arrangement of bulb turbine

The water flows axially towards the unit in the center of the water conduit and passes the generator, the main stays, the guide vanes, runner and draft tube and enters into tailrace channel.

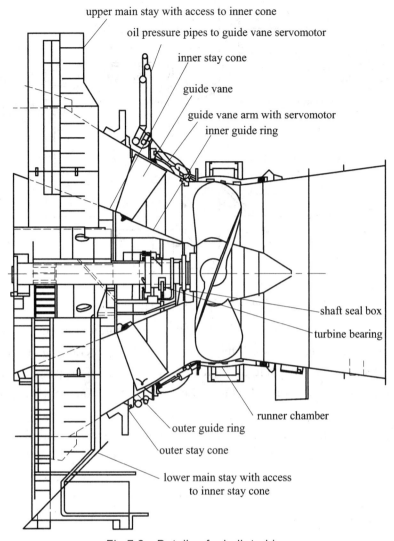

Fig.7.2 Details of a bulb turbine

7.3 Main Components

The bulb turbine consists of the following main components: stay cone, runner chamber, draft tube cone, generator hatch, stay shield, rotating parts, turbine bearing, shaft seal box, and guide vane mechanism.

7.3.1 Stay Cone

A longitudinal section through a stay cone structure and draft tube is shown in Fig.7.3. There are one lower (1) and one upper (5) main stay. An inner stay cone (3) and outer stay cone (6) are welded to the main stays. To the downstream end of the inner stay cone, the inner guide ring is bolted as shown in Fig.7.2.

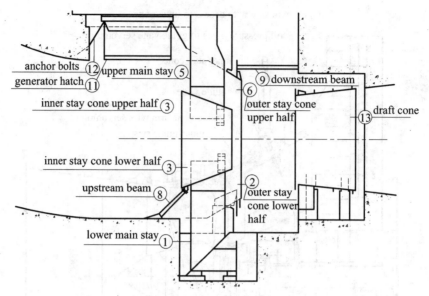

Fig.7.3 Vertical section of stay cone and draft tube

This outer stay cone forms a part of the outer water conduit contour and is embedded in concrete together with outer parts of the main stays. The generator bulb is bolted to the upstream end of the inner stay cone as indicated in Fig.7.1. These parts are located in the center of the water flow and forms the inner water conduit contour together with the runner hub.

Two side stays (4) and (7) in Fig.7.4, are located on each side of the bulb upstream of the main stays for stiffening the bulb and avoiding resonant vibrations.

The total weight and the hydraulic forces are transferred to the surrounding concrete through the stay cone via the two main stay vane structures. The dynamic as well as the static forces from the turbine and the generator are transferred through the structure to the building foundation.

7.3.2 Runner Chamber and Draft Tube Cone

The runner chamber is the connecting part between the outer stay cone and the draft tube cone (Fig.7.2). The downstream end of the outer cone is provided with a flange to which the runner chamber is bolted.

The draft tube cone consists of two or more straight welded steel cones and is embedded in concrete. The upstream end is connected to the runner chamber through a flexible telescope connection. This type of connection allows a certain axial movement of the runner chamber and the outer guide ring in order to withstand elongation/contraction due to temperature changes.

The length of the steel cone lining is determined by requirements to maximum water velocity at the exit and to avoid damage to the concrete.

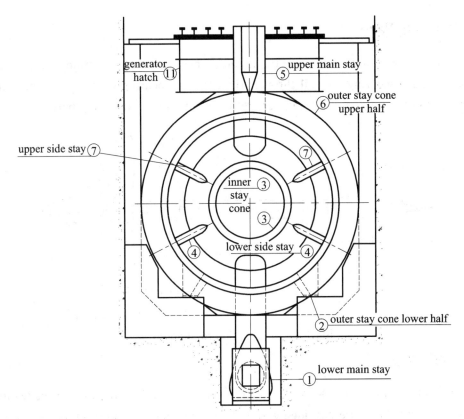

Fig.7.4 Cross section of stay cone

7.3.3 Generator Hatch

The generator hatch (11) in Fig.7.4 is normally a part of the turbine delivery. It is located above the generator and provides access to the generator for assembly or dismantling tasks.

The hatch consists of a perforated part which forms the outer water conduit contour in the hatch opening. A cylindrical steel mantel with a flange on top is provided for the hatch cover and seal mounting. As the unit's bulb part will rise and lower with filling and draining of the turbine, the seal joint between mantel and hatch cover must allow a vertical movement of the mantel.

7.3.4 Stay Shield

The stay shields are located between the generator access shaft and the turbine main stay. They form an even wall for the water flow and the stay structure is streamlined at the upstream end to prevent undesired vortex formation. The shields are bolted to the bulb and connected to each other by screw stays for stiffness. They are freely supported against the access way and the main stay to allow the axial movements.

The shields are provided with a manhole for inspection and possible maintenance of the space between them.

7.3.5 Rotating Parts

The rotating parts are shown in Fig.7.5 and consist of: runner; turbine shaft; shaft seal cam, clamp ring, wear ring and oil thrower ring; turbine bearing oil thrower ring; feedback mechanism and oil piping; and oil transfer unit from rotating to stationary parts.

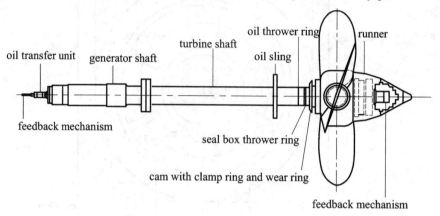

Fig.7.5　Bulb turbine rotating parts

7.3.5.1　The Runner

The runner is similar to a Kaplan runner and has normally three to five blades made of stainless steel. The blades are designed with flanges and connected to trunnions and levers.

The servomotor for moving the blades is normally located inside the hub as shown in Fig.7.6. The servomotor consists of a fixed piston and an axially moving cylinder, and link supports, links and blade levers are located inside the hub.

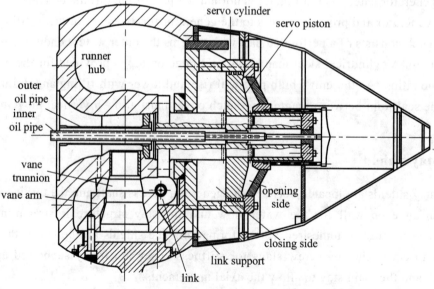

Fig.7.6　Runner hub

A photo of a bulb turbine runner is shown in Fig.7.7.

Fig.7.7 Runner assembly in the workshop

7.3.5.2 The Turbine Shaft

The turbine shaft is made of forged Siemens Martin steel and has flanges at both ends. One end is connected to the runner hub and the other to the generator shaft. These joints are pure friction joints.

7.3.6 Shaft Seal Box

Several types of shaft seal boxes are in use. One type as shown in Fig.7.8 is, however, used especially for the bulb turbines.

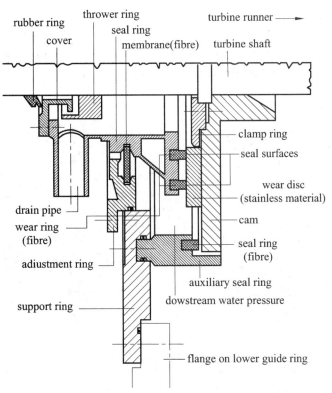

Fig.7.8 Shaft seal box

This box has radial seal surfaces consisting of a stainless hardened wear disk and two wear rings made of teflon type fibers. The wear disk is bolted to a cam fixed to the shaft. The wear rings are glued to the seal ring. This is movable and supported in the adjustment ring by means of a membrane.

The membrane allows the seal ring to move axially by 5-6 mm. This is necessary for the shaft movement in the downstream direction when the unit is loaded. In addition allowance must be made for wear of the seal surfaces.

The adjustment ring is bolted to the support ring and may be axially adjusted by means of double acting jacking screws. According to the wear range of wear rings, the adjustment range of the seal box should be 8-10 mm.

The auxiliary seal is located in the support ring. This can be pushed against or pulled away from the cam by means of push/pull jacking screws. When this ring is in contact with the cam, the wear seal rings can be dismantled without draining the unit.

Possible water leakage into the seal box is drained through a pipe to the pump sump.

A thrower ring is mounted on the shaft to prevent water leakage along the shaft. A rubber ring is mounted on the upstream end of the shaft and seals against the seal box cover.

The seal box is provided with four springs which press the wear seal rings against the seal surface to prevent leakage when the balance system is out of operation, e.g., when the turbine is filled.

7.3.7 Turbine Bearing

An example of a bearing design is shown in Fig.7.9. The bearing is sturdy and simple in operation. Maintenance will normally consist of oil change only.

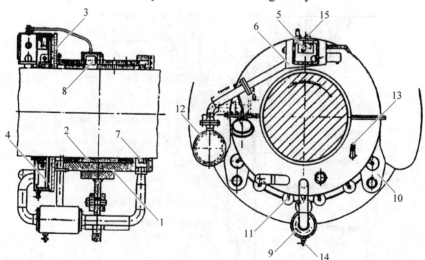

Fig.7.9 Turbine bearing

1—bearing housing; 2—bearing pad; 3—oil housing; 4—rotating oil reservoir; 5—oil scoop; 6—float box; 7—oil thrower ring; 8—automatic start lubrication unit; 9—sediment collector; 10—fork; 11—wedge; 12—oil tank; 13—filling opening; 14—drain valve; 15—inspection filling plug

The bearing housing (1) is supported in the inner guide ring (Fig.7.2), by means of two yokes and two support stays and rests normally on six wedges. By moving these wedges axially the bearing housing can be vertically adjusted. The bearing housing is split horizontally.

The bearing pad shell (2) (Fig.7.9) consists of an upper and a lower part. These are "floating" in the bearing housing where a radial locking pin in the upper part prevents the shells from rotating. The surface of the lower supporting shell and a surface in each end of the upper bearing shell are lined with babbit metal.

The oil housing (3) is bolted to the upstream end of the bearing housing. The oil reservoir (4) is fixed to the shaft. The oil scoop (5) and the box (6) are also located inside the oil housing. The float box is provided with a window for observation of the oil circulation when the turbine is running.

An oil thrower ring (7) is fixed to the shaft to prevent oil from seeping out of the bearing along the shaft in the downstream end.

The automatic start lubrication unit (8) is mounted on the top of the bearing housing. This consists of a container which is filled with oil when the turbine is running. When the shaft stops, the oil content is kept in the container by a support device held by the shaft. As soon as the shaft resumes rotation, the support is removed and the content is tilted. The oil in the container is then distributed on the bearing surface.

When the shaft and oil reservoir start to rotate, the reservoir draws oil from the lower half of the oil housing. As soon as the oil layer is sufficiently thick, the oil scoop picks up oil and delivers it to the float box, then to the oil tank and the bearing pad. The rotating shaft transports this oil further to the bearing surface.

Normally more oil than required for lubrication is circulating through the oil tank. Therefore, a bypass is provided for taking the excess oil back to the oil housing top. This bypass flow is controlled by a float switch inside the float box (6).

To increase the oil volume to more than the volume of the oil housing, an oil tank (12) is located beside the bearing.

The sediment collector (9) is located below the bearing housing. All dirt particles, which are trapped in the oil during circulation in the bearing, shall be separated before the oil returns to the oil housing.

The bearing is provided with miscellaneous filling openings, oil level indicators, level float and temperature sensors as well.

7.3.8 The Feedback Mechanism and Oil Piping

The feedback mechanism and oil transfer piping are located in the shaft center. The transfer piping consists of an inner and an outer concentric oil pipe running through the whole shaft length. The inner pipe continues through the oil transfer unit at the upstream end, and it is supported in the outer pipe and connected to the runner servomotor cylinder via an yoke.

The inner pipe is axially movable and follows the servomotor movement. A pointer mounted to the upstream end moves along a measurement ruler showing mechanically the servomotor position at any time. The outer oil pipe is mounted by means of flange connections to the runner hub, turbine shaft and generator shaft respectively.

7.3.9 The Oil Transfer Unit

The oil transfer unit is located at the upstream end of the generator shaft and has one fixed and one rotating part consisting of a distribution sleeve and distribution trunnion respectively.

The distribution sleeve is fixed to the capsule around the generator shaft end and is provided with pipe connections for oil supply and return as well as leakage oil. The distribution sleeve is provided with a bracket with a measurement scale where the runner servomotor position can be read.

7.3.10 The Guide Vane Mechanism

Two different systems have been used for operating the guide vanes. Kværner Brug has designed a system where each particular vane has its own servomotor as shown in Fig.7.10.

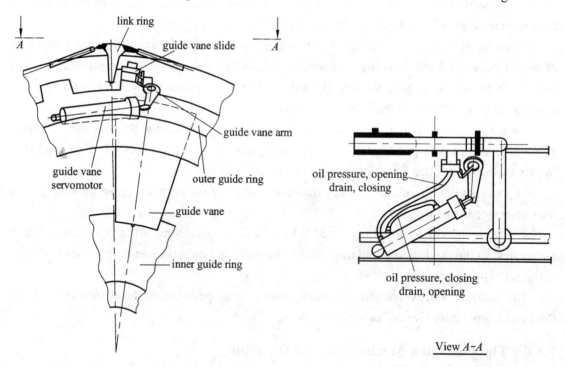

Fig.7.10 Guide vane mechanism with single servomotors

By means of a link ring, a simultaneous movement of the pilot valves for the guide vane servomotors is achieved. The movement is governed by the valves controlling the opening/closing of the guide vanes.

How the servomotors are supplied with or drained for oil is shown in section *A-A* in Fig.7.10.

The high pressure hoses are connected to the oil pressure system of the unit.

The advantage of this system is that even if one guide vane is stuck, the remaining vanes can be moved without any damage. The same will apply if a foreign object is caught and jammed between two vanes during closing, the remaining vanes can be closed without any damage. If required, single vanes may be operated separately and thus jammed object may easily be flushed out of the guide vane system.

The disadvantages of this system are the assembly and extensive adjustment work. However, the total price will approximately be the same as for the regulating ring system which is the other method for moving the guide vanes.

The regulating ring system with links and levers is of the same type as for Francis turbines. A regulating ring with three main servomotors is shown in Fig.7.11. Due to the conical arrangement of the guide vanes, the lever link system must have spherical bearings with large angle movement.

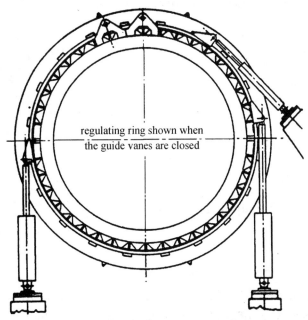

Fig.7.11 Regulating ring

The connection between the levers and the vanes is designed as friction joints. This is done to avoid damage to parts if one or several vanes are stuck or if foreign objects are caught between vanes.

The friction joint makes it possible for the vane lever to move with the remaining parts of the guide vanes connected to the regulating ring even if the adjacent vane is stuck.

A disadvantage may be that large weight of the regulating ring and possible slack in bearings can make the governing inaccurate.

7.4 Assembly and Dismantling

Among the large main parts of the bulb turbine, it is only the runner that is normally needed to dismantle. The runner chamber is split axially in two horizontal halves. By removing the upper half access to the runner is obtained.

The guide vanes cannot be dismantled without extensive work. Repairs of these and the guide surfaces should be performed at the plant.

Bearing and seal box can easily be dismantled. By applying the overhaul seal, the seal box may be removed without draining the water canal. Then necessary stairs and floors around the guide vane and the runner chamber can be erected.

The stay shields are adapted against bulb and outer water conduit contour. The shields are mounted as soon as the generator bulb and penstock are completed.

Finally the generator hatch dome plate and cover are installed.

Questions

1. Given a drawing of a bulb turbine, identify the following major components:
   - Generator hatch
   - Bulb
   - Access to turbine
   - Access to generator
   - Stay cone
   - Guide vanes
   - Guide ring
   - Runner
   - Draft tube

2. State the purposes of the following components in a bulb turbine:
   - Stay cone
   - Stay shield
   - Feedback mechanism and oil piping

3. Explain the structure of the shaft seal box and the turbine bearing.

4. Illustrate the two types of guide vane mechanisms used in bulb turbines.

References

[1] Arne Kjølle. Hydropower in Norway-Mechanical Equipment [M]. Trondheim: Norwegian University of Science and Technology, 2001.

[2] GE Renewable Energy. Bulb Hydro Turbine [EB/OL]. [2016-06-16].https://renewables.gepower.com/hydro-power/large-hydropower-solutions/hydro-turbines/bulb-turbine.html.

[3] 国家能源局. 灯泡贯流式水轮发电机组安装工艺规程: DL/T 5038-2012 [S/OL]. 北

京：中国电力出版社，2012：27-140 [2016-06-16]. http://www.doc88.com/p-9783657329240.html.

[4] 田树棠. 贯流式水轮发电机组及其选择方法[M]. 北京：中国电力出版社, 2000.

Important Words and Phrases

access　*n.* 入口；通道；
　　　　v. 接近，进入；使用接近，获取
access shaft　入口井；竖井通道；进口竖井
adjustment ring　调整环
allowance　余量；限额
anchor bolt　锚固螺栓；地脚螺栓
assembly　装配；组装
babbit metal　巴比特合金
bearing housing　轴承座
bearing pad　轴承垫片
blade lever　叶片转臂
bracket　支架；托架；括号
bypass　旁通管；旁通；旁路
capsule　封装体
clamp ring　锁紧圈
concentric　同轴的；同心的
container　容器；集装箱
contour　外形，轮廓
contraction　收缩，紧缩
cylinder　圆柱；圆柱体；油缸
delivery　配送；传送，投递；交付
dismantling　拆开；拆卸
dome plate　顶板
double acting jacking screw　双动式顶起螺丝
draft tube cone　尾水锥
drain valve　放空阀
drain　使流出；排掉
elongation　伸长；延长
embed　嵌入；埋入
even　平坦的；相等的
feedback mechanism　反馈机制，回馈机制
filling opening　注油孔
float box　浮动框

float switch　浮动开关
floating　不固定的，流动的，浮动的
flush out　冲掉，排出
foreign object　异物；外物
forged　锻的；锻造的
fork　U字形辊架
friction joint　摩擦接头
generator hatch　发电机舱
guide ring　配水环
guide vane arm　导叶臂
halve　二等分
hardened　硬化的；淬火的
hose　软管
inspection　检修；检查
jacking screw　顶起螺丝；顶起螺钉；螺旋千斤顶
jam　使堵塞；挤进，使塞满
leakage oil　渗漏油
level float　油位浮标
link ring　连接环
link support　支撑连杆
locking pin　定位销钉；锁定销
main stay　主支柱
manhole　人孔，检修孔；探孔；入孔；进人孔
mantel　盖板；壁炉架
measurement scale　量尺；测量尺度
membrane　隔膜
miscellaneous　各种各样的
oil housing　集油罩
oil layer　油膜
oil level indicator　油位指示器
oil piping　润滑油管系；油管
oil pressure system　油压系统

oil reservoir 储油器	sleeve 套筒，套管
oil scoop 集油盘	spherical bearing 球面轴承
oil sling 油管吊索	spindle 轴
oil tank 油箱	spring 弹簧
oil thrower ring 甩油环；甩油圈	stay cone 座环
oil transfer unit 受油器	stay shield 导流板
overhaul 大修；检修	stiffen 使变硬；加强
perforated 穿孔的	streamlined 流线型的；改进的
pilot valve 操纵阀	stuck 被卡住的；不能动的
pointer 指针	sturdy 坚固的
pump sump 废水收集池；集水坑	support ring 支撑环
resonant vibration 共振；谐振	support stay 撑杆
resume 重新开始；恢复	tailrace channel 尾水渠
rubber ring 橡胶圈	teflon type fiber 聚四氟乙烯纤维
screw stay 螺旋撑条；撑条	telescope connection 伸缩管式连接；伸缩节
seal joint 密封接头	temperature sensor 温度传感器
seal surface 密封面；密封表面	tilt 倾斜；翘起
sediment collector 沉积收集器	vortex 涡流；漩涡
seep out 渗出	water conduit 水管
shaft seal box 轴封盒	wear disk 耐磨盘
side stay 侧支柱	wear ring 耐磨环；磨损环；磨耗环
Siemens Martin steel 平炉钢	wedge 楔块
slack 松弛的；不流畅的	yoke 轭架

8 Turbine Governing

8.1 Introduction

When a turbine drives an electrical generator or alternator, the primary requirement is that the rotational speed of the shaft and hence that of the turbine rotor has to be kept fixed. Otherwise the frequency of the electrical output will be altered. But when the electrical load changes depending upon the demand, the speed of the turbine changes automatically. This is because the external resisting torque on the shaft is altered while the driving torque due to the change of momentum in the flow of fluid through the turbine remains the same. For example, when the load is increased, the speed of the turbine decreases and vice versa. A constancy in speed is therefore maintained by adjusting the rate of energy input to the turbine accordingly. This is usually accomplished by changing the rate of fluid flow through the turbine – the flow is increased when the load is increased and the flow is decreased when the load is decreased. This adjustment of flow with the load is known as the governing of turbines.

The regulation of speed is normally accomplished through flow control. A Francis turbine is controlled by opening and closing the wicket gates, which varies the flow of water according to the load. The actuator components of a governor are required to overcome the hydraulic and frictional forces and to maintain the wicket gates in fixed position under steady load. For this reason, most governors have hydraulic actuators. The Kaplan turbine is characterized by that both the wicket gates and the turbine blades are adjustable. The regulation of the two are coordinated and combined to gain maximum efficiency. For every wicket gate position, the turbine blade position is chosen such that it maximizes the ratio of output power and water volume. On the other hand, impulse turbines are more easily controlled. This is due to the fact that the jet can be deflected or an auxiliary jet can bypass flow from the power-producing jet without changing the flow rate in the penstock. This permits long delay time for adjusting the flow rate to the new power conditions. The spear or needle valve controlling the flow rate can close quite slowly, say, in 30 to 60 seconds, thereby minimizing any pressure rise in the penstock.

8.2 Turbine Governing Demands

8.2.1 Frequency and Load Regulation

The governor shall be able to maintain stability of the generating unit when running on an isolated grid. Generally the units are designed for stable operation up to full load. In this mode of operation the governor shall keep the frequency within certain limits of deviation.

Load regulation on a rigid system is the most common operation mode. Each unit has little influence on the grid system frequency. The governor controls the load to the desired value. The variation of the load as a function of the change in frequency is dependent on the permanent droop setting.

A special mode of operation is the manual mode where the guide vane openings are controlled manually by means of a mechanical hydraulic load limiter. In this mode only the load can be controlled.

8.2.2 Start and Stop Sequence Control

During the period of start, the unit shall be run up to nominal speed as quickly and smoothly as possible. The start can be carried out both manually and automatically. The admission must be opened only when permitted by all overriding start conditions.

In shut down mode, the admission shall be closed as quickly as possible but limited by the magnitude of the pressure rise in the tunnel and pressure shaft system. Due to safety reasons, the shut down signal will be given simultaneously to different stages in the governor, e.g., closing of the load limiter or the emergency operated shut down valve. The shut down valve also functions if the ordinary voltage supply has failed. The stop command can be given both manually and automatically.

8.2.3 Disconnection, Load Rejection

Disconnection means to open the generator main circuit switch. The generator is thereby separated from the grid and the turbine power output results in a speed rise of the unit. The function of the governor is then to shut down the turbine not too fast to avoid pressure rise above the guaranteed level.

8.2.4 Load Limiting

Load limiting must be possible according to external conditions. The load limiter device may be operated both manually and automatically.

8.3 Functions of Governors

Much equipment connected to a hydroelectric system is sensitive to frequency variation. Therefore, speed control of the system is a necessity. Regulating the quantity of water admitted to the turbine runner is the usual means of regulating and maintaining a constant speed to drive the generator and to regulate the power output. This is done by operating wicket gates or valves. Such action requires a mechanism to control the wicket gates, which is the governor or governor system. Turbine governors are equipment for the control and adjustment of the turbine power output and evening out deviations between the power and the grid load as fast as possible.

The turbine governors have to comply with two major purposes:

① To keep the rotational speed stable and constant of the turbine-generator unit at any grid load and prevailing conditions in the water conduit.

② At load rejections or emergency stops the turbine admission have to be closed down according to acceptable limits of the rotational speed rise of the unit and the pressure rise in the water conduit.

Alterations of the grid load cause deviations between the turbine power output and the load. For a load decrease, the excess power accelerates the rotating masses of the unit according to a higher rotational speed. The following governor reduction of the turbine admission means deceleration of the water masses in the conduit and a corresponding pressure rise.

The governing of water turbines requires limitations of speed rise and pressure rise as well as fulfillment of regulating stability demands. To keep the rise of the rotational speed below a prescribed limit at load rejections, the admission closing rate must be equal to or higher than a certain value. For the pressure rise in the water conduit the condition is opposite, e.g., the closing rate of the admission must be equal to or lower than a certain value to keep the pressure rise as low as prescribed.

For power plants where these two demands are not fulfilled by one single control, the governors are provided with dual control functions, one for controlling the rotational speed rise and the other for controlling the pressure rise. This is normal for governors of high head Pelton and Francis turbines.

For Pelton turbines the principle is:

—To set the closing rate of the needle control of the nozzles to a value which satisfies the prescribed pressure rise.

—To bend the jet flow temporarily away from the runner by a deflector so the speed rise does not exceed the accepted level.

For Francis turbines the principle is:

—To set the closing rate of the guide vane opening to a value, which satisfies the rotational speed rise limits.

—To divert as much of the discharge through a controlled bypass valve that the pressure rise in the conduit is kept below the prescribed level.

8.4 Specific Turbine Governing Equipment

8.4.1 Dual Control of Pelton Turbines

The governing of Pelton turbines is normally carried out with a dual control, e.g., needles in the nozzles and the deflectors.

During minor load changes the needle adjustment control can satisfy the control requirements alone. During rapid load rejections, however, the rotational speed rise has to be controlled and limited by activation of the deflector. The servomotor gives the deflector a rotary movement which bends the water jet away from the runner.

The sequence controlled nozzle follows the movement of the deflector servomotor by adjustment of the needle position until the discharge corresponds to the new power/load equilibrium. In this approach to equilibrium, the deflector moves gradually out of the jet again to an idle position just outside the periphery of the jet.

The deflector function is controlled from the governor desk via the main control valve. The servomotor movement is transferred via bars, levers and links to the deflectors.

The servomotor movement feedback signal is transferred to the control valve.

The sequence control of each needle is carried out via a cam disk which is driven by the deflector servomotor. The cam shape accomplishes an input function to the needle control valve which opens for adjusting the needle to correct position.

The feedback from the needle movement controls the valve openings until they reach neutral position and the needle has reached the correct position according to the reference.

The cabinet for the needle control is located as near to the nozzles as possible to obtain a plain feedback system from the needles and short oil pressure pipelines.

In some recent designs of Pelton governors electronics and separate electrohydraulic servosystems carry out the controller functions of the deflectors and the needles. Electronics is also utilized in the feedback systems. In this way the construction of the servomotors with mechanical levers and links has become more simplified.

8.4.2　Bypass Control for Francis Turbines

8.4.2.1　Function and General Arrangement

The regulation following load rejections in high head Francis turbine plants makes it necessary to divert some parts of the instantaneous water flow from the turbine. In this way it is possible to obtain a rapid closure of the turbine admission and to retard the main flow in the penstock to minimize the pressure rise.

The diversion is normally branched off from the scroll casing as shown schematically in Fig.8.1. In this branched-off system the bypass valve is installed and its admission is controlled by the turbine governor. The guide vane movement controls the valve opening. This combined control of the guide vane movement and the bypass through the valve makes it possible to control both the pressure rise and speed rise during load rejection.

The water flow through the valve is led into an energy dissipater and then into the turbine draft tube as shown in Fig.8.2.

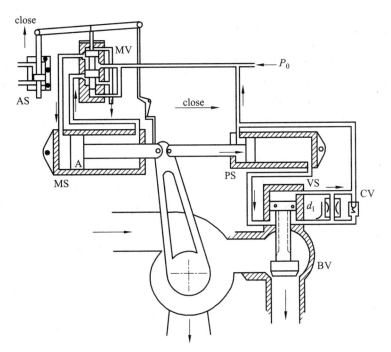

Fig.8.1 Bypass valve with regulating system

AS—auxiliary servomotor; MV—main valve; MS—main servomotor; PS—pilot servomotor; VS—valve servomotor; BV—bypass valve; CV—check valve; d_1—orifice for speed control; P_0—oil pressure

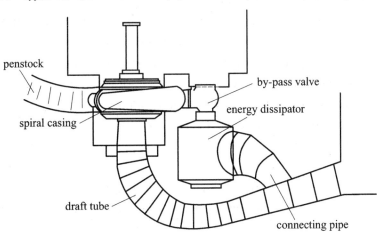

Fig.8.2 Energy dissipater

A Norwegian turbine manufacturer has used this type of bypass valve system for several years, and a long experience in the use of this equipment has proved a simple and reliable system.

A vital point in the design of the control system is to ensure full control of the pressure rise even if the valve should fail to operate. In this case the closing speed of the guide vanes will correspond to the speed given by the allowable pressure rise. The speed will then be higher than normal, but this is regarded as less serious because the generator is designed to withstand short-lasting runaway speed of the unit.

8.4.2.2 The Valve Control System

A schematic diagram of the valve system and main servomotor control is shown in Fig.8.1. The auxiliary servomotor AS, via the main valve MV controls the guide vane servomotor MS. The double acting servomotor MS, moves the guide vane via the regulating ring. A pilot servomotor PS, is connected to this ring and copies the position according to the movement of MS.

The servomotor PS is pressurized via the orifice d_1 from oil pressure supply P_0. The bypass valve servomotor VS is hydraulically connected to PS. Under stationary conditions it moves to closed position because of the difference between the two sides of the piston areas.

During a closing movement of MS the oil from PS passes through d_1 into the accumulator. If the closing speed exceeds a certain value, the pressure on the opening area on VS increases because of the orifice d_1, and VS then opens.

To avoid restriction of the guide vanes a check valve CV, is connected in parallel to d_1.

The size of the volume of VS relative to PS is given by the size of the opening of the bypass valve at total flow capacity.

8.4.3 Dual Control of Kaplan/Bulb Turbines

The optimal efficiencies of Kaplan and bulb turbines are obtained by the optimal combination of the functions of the guide vane cascade control and the runner blade control.

The combination of the two control functions may be carried out either by mechanical-hydraulic or by electrohydraulic operation. The combination unit is usually located on the top or beside the top of the turbine-generator unit.

A mechanical-hydraulic combination unit is integrated in the runner blade control system and consists of main valve, feedback mechanism, combination control function curve, and pipe connections to the oil supply unit.

The combination control function is to distribute oil for operation of the runner blades via the main valve, to position the runner blades according to the control function curve disk which is governed by the guide vane control, and to feedback of the spool position in the main valve.

An electrohydraulic combination unit has an electrical feedback from the position of the guide vane cascade and a combination control function in this case is produced electronically. The servomotor is then operated by an electrohydraulic servovalve.

Questions

1. Explain the necessity of turbine governing.
2. List the purposes of turbine governors.
3. Explain the governing principles of Pelton turbines and Francis turbines.
4. Illustrate the bypass valve control system of Francis turbines.

References

[1] Arne Kjølle. Hydropower in Norway-Mechanical Equipment[M]. Trondheim: Norwegian University of Science and Technology, 2001.

[2] Warnick C C. Hydropower engineering[M]. New Jersey: Prenticc-Hall, Inc., Englewood Cliffs, 1984.

[3] The World Bank. Chapter 4: Electro Mechanical Equipment [EB/OL]. [2016-06-16]. http://wbi.worldbank.org/energy/small-hydropower-technology/chapter-4-electro-mechanical-equipment.

Important Words and Phrases

accumulator　收集器
actuator　执行机构
alter　改变，更改
alteration　改变；变更
alternator　交流发电机
auxiliary servomotor　辅助接力器
cam disk　凸轮盘
deviation　偏离；偏差；误差
disconnection　断开；切断
double acting　双动式
driving torque　驱动转矩；传动转矩；传动力矩
dual control　双重控制
electrical load　电力负荷；电力负载
electrical output　电力输出
electrohydraulic　电动液压的
electronics　电子器件
energy dissipater　消能器
equilibrium　平衡
even out　使均等

feedback signal　反馈信号
frequency　频率
generating unit　发电机组
governing　调速；控制，调节
grid　电网
idle position　停止位置
load limiter　载荷限制器；负荷限制器
load regulation　负荷调整
main circuit switch　主电路开关
main servomotor　主接力器
orifice　孔口
overriding　过载
permanent droop　永态转差率
pilot servomotor　中间接力器
prescribed　规定的
resisting torque　阻力矩；抗力矩；抗转矩
retard　延迟；使减速
servovalve　伺服阀
spool position　阀芯位置
turbine governing　水轮机调速

9　Selection of Hydraulic Turbines

Every hydropower plant is unique. Comparing one hydropower plant to the other is like comparing orange with apple. To harness water discharge of a river to the maximum, the selection of hydro turbine is very important. A wrong selection of hydro turbine can change the whole salient feature of a hydropower plant.

Once a hydro power project site's head, flow rate, and power requirements are determined, we can use Fig.9.1 to assist in determining which turbine suits the situation. The flow rate Q is plotted on the horizontal axis and is expressed in terms of m³/s. The net head available H is plotted on the vertical axis and is expressed in terms of m. The power output P is plotted on a third axis at 45 degrees to the flow and the head axes and is expressed in terms of MW.

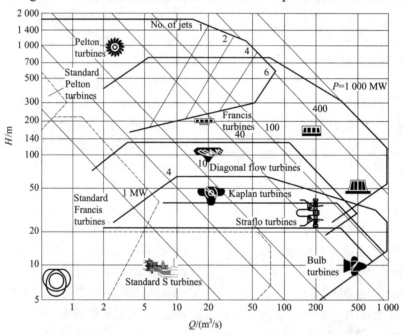

Fig.9.1　Overview of turbine runners and their operating regimes

The power that can be developed by a turbine is a function of both the head and the flow available:

$$P = \eta \rho g Q H \tag{9.1}$$

where

　　η—the turbine efficiency;
　　ρ—the density of water (kg/m³);
　　g—the acceleration due to gravity (m/s²);
　　Q—the flow rate (m³/s);
　　H—the net head (m).

The net head available to the turbine dictates the selection of type of turbine suitable for use at a particular site. The rate of flow determines the capacity of the turbine. The term specific speed is generally used in classifying the types of turbines and in defining the characteristics within each type.

The selection of turbines depends upon the following factors:
- Site characteristics;
- Head of the hydro scheme;
- Flow rate available in the scheme;
- Desired runner speed of the generator;
- The probability of operating the turbine at reduced flow rates.

9.1 Speed

To develop a given power at a specified head for the lowest possible first cost, the turbine and generator unit should have the highest speed practicable. However, the speed may be limited by mechanical design, cavitation tendency, vibration, drop in peak efficiency, or loss of overall efficiency because the best efficiency range of the power efficiency curve is narrowed. The greater speed also reduces the head range under which the turbine will satisfactorily operate.

9.2 Specific Speed

The specific speed is generally the basis of the selection procedure. This term is specified as the speed in revolutions per minute at which the given turbine would rotate, if reduced homologically in size, so that it would develop one metric horsepower at full gate opening under one meter head. Low specific speeds are associated with high heads and high specific speeds are associated with low heads, as shown in Fig.9.2. Moreover, there is a wide range of specific speeds which may be suitable for a given head.

The specific speed (metric HP units) range of different types of turbines is as follows:
- Fixed blade propeller turbines: 300-1,000.
- Adjustable blade Kaplan turbines: 300-1,000.
- Francis turbines: 65-445.
- Pelton turbines: 16-20 per jet. For multiple jets the power is proportionally increased.
- Crossflow turbines: 12-80.

Selection of a high specific speed for a given head will result in a smaller turbine and generator, with savings in capital cost. However, the reaction turbine will have to be placed lower, for which the cost may offset the savings. The values of electrical energy, plant factor, interest rate, and period of analysis are used in the selection of an economic specific speed. Commonly used mathematical expression for specific speed is power based (English System), as follows:

$$n_s = \frac{n_r \sqrt{P_r}}{H_r^{5/4}} \qquad (9.2)$$

where

n_r—rotational speed of the turbine in revolutions per minute;
P_r—power in metric horsepower at full gate opening (1 kW = 0.86 metric hp);
H_r—rated head in m.

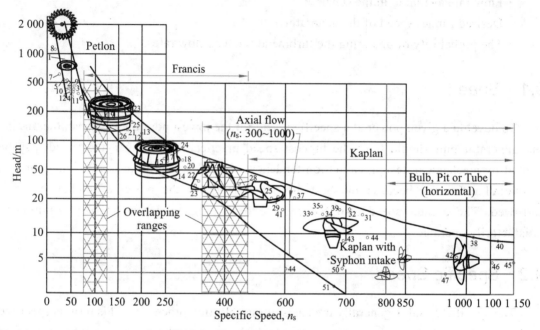

Fig.9.2 Specific speed versus Head

The specific speed value defines the approximate head range application for each turbine type and size. Low head units tend to have a high specific speed, and high-head units to have a low specific speed. Once the specific speed is determined, Fig.9.2 can be used to determine the type of turbine that may be adopted for a particular project. Impulse turbines are efficient over a relatively narrow range of specific speed, whereas Francis and propeller turbines have a wider useful range.

Note in Fig.9.2 that there is an overlap in the range of application for various types of equipments. This means that either type of unit can be designed for good efficiency in this range, but other factors, such as generator speed and cavitation, may dictate the final selection.

9.3 Turbine Efficiency

An understanding of how the efficiency of turbines varies is useful in the selection of hydropower units. Typical efficiency curves of the various types of turbines are shown for comparison in Fig.9.3.

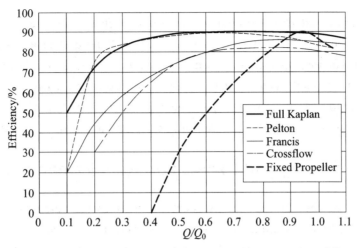

Fig.9.3 Efficiencies comparison of different turbines at reduced flow rates

These curves are shown to illustrate the variation in efficiency of the turbine through the load range of the design head. The relative efficiencies of different turbines can be compared at the design point and at the reduced flow rates. An important consideration is whether or not a turbine is required to operate over a wide range of loads.

9.3.1 Impulse Turbines

Impulse turbines normally have most economical application at heads above 1,000 ft, but for small units and cases where surge protection is important, impulse turbines are used with lower heads. Pelton wheels tend to operate efficiently over a wide range of power loading because of their nozzle design, but the efficiency of crossflow turbines falls sharply as they are operated below half of their design flow.

9.3.2 Reaction Turbines

In the case of reaction machines that have fixed geometry, such as Francis turbines, efficiency can vary widely with load. For Francis turbines the units can be operated over a range of flows from approximately 50% to 115% best efficiency discharge. Below 40%, low efficiency, instability, and rough operation may make extended operation unwise. The upper range of flow may be limited by the instability or the generator rating and the temperature rise. The approximate limits of head range from 60% to 125% of the design head.

9.3.3 Axial Flow Turbines

Axial flow turbines have been developed for heads from 5 to 200 ft but are normally used for heads less than 100 ft. Just like Francis turbines, fixed blade propeller turbines also have fixed geometry. Due to the large variation of efficiency with load, the limits of flow operation should be between 75% and 100% of the best efficiency flow.

However, Kaplan and Deriaz turbines can maintain high efficiency over a wide range of operating conditions, and their efficiencies are high when running below the design flow. Kaplan units may be operated between 25% and 125% of the best efficiency discharge. The head range for satisfactory operation is from 20% to 140% of the design head.

The decision of whether to select a simple configuration with a relatively "peaky" efficiency curve or to install a more complex machine with a broad efficiency curve but added expense will depend on the expected operation of the plant and other economic factors.

9.4 Turbine Setting and Excavation Requirements

The location of the turbine with respect to the tailwater elevation is an important consideration in turbine selection and installation. Turbines are subject to pitting due to cavitation, which is directly related to the setting. To limit cavitation, the turbine is normally set deeper, which increases the pressure on the turbine runner and minimizes the formation of vapor bubbles. Obviously, a deeper turbine setting results in increased civil costs for the plant construction. Higher turbine speed also increases the tendency for a turbine to cavitate, but higher turbine speed results in smaller equipment and a lower cost unit. The setting depth is in general a balance between equipment cost and civil costs.

To determine the cavitation performance of a turbine, we must consider the vapor pressure of the water, the atmospheric pressure, the setting of the runner with respect to the tailwater, and the maximum head at which the turbine operates at full gate. These factors are expressed by the Thoma Cavitation Coefficient in the following relationship:

$$\sigma = \frac{H_a - H_v - H_s}{H} \tag{9.3}$$

where

σ —Thoma Cavitation Coefficient;

H_a —atmospheric pressure in feet of water (see Table 9.1);

H_v —vapor pressure in feet of water (see Table 9.2);

H_s —elevation of the runner tailwater in feet measured at the throat of a Francis turbine, at the centerline of the blades for a vertical propeller turbine, or at the tip of the runner for a horizontal turbine (see Fig.9.4);

H —total head on the turbine at full gate.

Table 9.1 Atmospheric pressure at various altitudes

Altitude (ft)	H_a (ft of H_2O)	Altitude (ft)	H_a (ft of H_2O)
0	33.959	5,000	28.25
500	33.35	5,500	27.73
1,000	32.75	6,000	27.21
1,500	32.16	6,500	26.70
2,000	31.57	7,000	26.20
2,500	31.00	7,500	25.71

Continue

Altitude (ft)	H_a (ft of H_2O)	Altitude (ft)	H_a (ft of H_2O)
3,000	30.43	8,000	25.22
3,500	29.88	8,500	24.74
4,000	29.33	9,000	24.27
4,500	28.79	9,500	23.81
		10,000	23.35

Table 9.2 Vapor pressure of water at various temperatures

Temperature (°F)	H_v (ft)	Temperature (°F)	H_v (ft)
40	0.28	60	0.59
50	0.41	70	0.84

The value of σ for any specific installation is only affected by the change of H_s or H, since H_a and H_v are relatively constant. The value of σ at which cavitation occurs is called the critical value and is denoted by σ_c. Calcualted σ values for an installation that are less than σ_c will result in cavitation in the turbine. The value of σ_c is determined by model tests of a particular runner design. In the absence of test data, the following formula can be used to calculate the minimum value of σ:

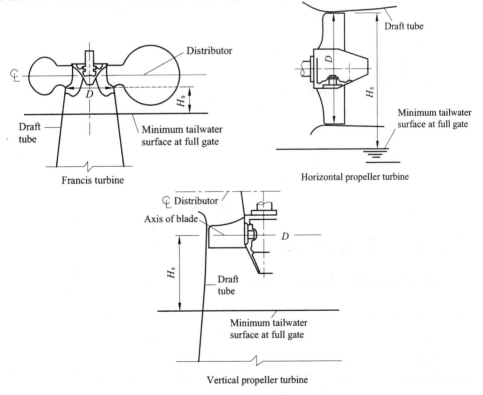

D—minimum turbine runner diameter; H_s—distance from minimum tailwater surface at turbine full gate to the minimum runner diameter for vertical units or to the tip of the blade for horizontal units

Fig.9.4 Turbine setting coefficient definition

$$\sigma = \frac{n_s^{1.6}}{4\,325} \tag{9.4}$$

Eq.9.4 is drawn graphically in Fig.9.5 so that calculation time can be saved. By rearranging Eq.9.3, the elevation of the runner at its throat can be calculated:

$$H_s = H_b - H_v - \sigma H \tag{9.5}$$

It is normal practice to subtract an additional distance to ensure that the runner is placed well below the cavitaiton limits. A rule of thumb for small units would be 3 feets. Eq.9.5 can now be rewritten in the following form:

$$H_s = H_b - H_v - \sigma H - 3 \tag{9.6}$$

This method should only be used for preliminary setting of the turbine since the setting will vary with different manufacturers. These calculations can be used during the initial project stages for design of the powerhouse, calculation of the needed excavation, and selection among different turbine types and designs.

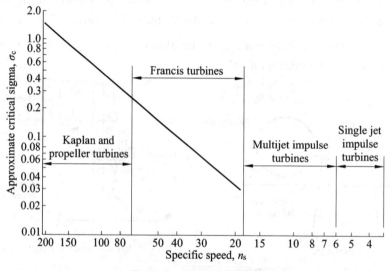

Fig.9.5 Critical sigma

9.5 Criteria for Selection of Hydraulic Turbines

9.5.1 General Guideline for Selection of Type of Turbine

⟡ Low head up to 20 m, discharge and load vary widely – Bulb turbine.

⟡ Medium and lower range of heads (up to 80 m) constant discharge – Fixed blade propeller turbine.

⟡ Medium and lower range of heads (up to 80 m) when discharge and load vary widely – Kaplan turbine.

⟡ Medium and high range of heads (30-550 m), turbine discharge varying in acceptance limits – Francis turbine.

◇ Head greater than 400 m, varying discharge and load conditions – Pelton turbine.

The factors given in the following table determine the type of turbine to be used depending upon site conditions.

Table 9.3 Application scope of different types of turbines

Type of turbine	Approx. head range (m)	Head variation (% of rated head)	Load variation (% of rated load)	Specific speed range (MHP)	Peak efficiency (percentage)
Pelton	>300	120-80	50-100	15-65	92
Francis	30-400	125-65	50-100	60-400	96
Deriaz	50-150	125-65	50-100	200-400	95
Kaplan	10-60	125-65	40-100	300-800	95
Propeller	10-60	110-90	90-100	300-800	93
Bulb	5-20	65-125	40-100	600-1200	95

9.5.2 Criteria for Selection of Hydraulic Turbine in Overlapping Zone

(1) Minimum output for continuous operation.

Table 9.4 Minimum output of different turbines

Type of turbine	Minimum output for continuous operation
Pelton	30
Francis	50
Kaplan/Bulb	30
Propeller	85
Deriaz	40

(2) Turbine setting and excavation requirements.

(3) Transport considerations (runner).

The difficulty of transporting large runners sometimes makes it necessary to limit their size. A runner with a maximum overall diameter of 18 ft (5.5 m) is about the largest that can be shipped by rail. Larger units require construction in segments and field fabrication with special care. Field fabrication is costly and practical only for multiple units where the cost of facilities can be spread over many units. Runner may be split in two pieces, completely machined in the factory and bolted together in the field. This is likewise costly, and most users avoid this method because the integrity of the runner cannot be assured.

(4) Pressure rise and speed rise considerations.

Table 9.5 Pressure rise and speed rise of different turbines

Type of turbine	Pressure rise (%)	Speed rise (%)
Pelton	15-20	20-25
Francis	30-35	35-55
Kaplan/Bulb & Propeller	30-50	30-65
Deriaz	20-45	35-65

(5) Turbine efficiency.

① The performance of a turbine is ideal at the design head. But the fall of efficiency in case of Pelton, Kaplan and Bulb turbines is much less in comparison to Francis and Propeller types. Therefore in overlapping head ranges selection of type of turbine should consider the head variation existing at site.

② Turbine efficiency varies with load. The fall of efficiency at part load for Francis and Propeller is much steeper in comparison to that for Kaplan and Pelton turbines. Therefore, necessity of operating turbines at part loads for longer time influences the choice of turbines in the overlapping head ranges.

Thus in the head ranges where both Kaplan and Francis are suitable, the requirement of large pressure head and electrical load variation dictates, Kaplan turbine to be superior to Francis turbine from considerations of higher power generation on account of better overall efficiency.

Similarly, in the overlapping head ranges where both Francis and Pelton could be used, Pelton has advantages over Francis in overall performance level when variation of load and head is higher.

③ The highest specific speed of a turbine resulting in a higher speed of rotation for a generator with consequent reduction in the cost of the generator. This criterion is very important for selecting the type of turbine from cost consideration in the overlapping head ranges.

(6) Susceptibility to silt.

9.6 Determination of the Number of Units

Normally, it is most cost effective to have a minimum number of units at a given installation. However, multiple units may be necessary to make the most efficient use of water where flow variation is great. Factors such as space limitations by geologic characteristics or existing structure may dictate larger or smaller units.

Fig.9.6 shows how multiple units can be used effectively to take advantage of low-flow variation. Two-unit, three-unit, and four-unit plants of equal unit capacity are preferred and should provide for any variation in flow. At the design stage of analysis and with the availability of standardized units, it may be desirable to consider as alternatives a single full-capacity unit, two or more equal-size units, and two or more unequal-size units to determine the optimum equipment selection.

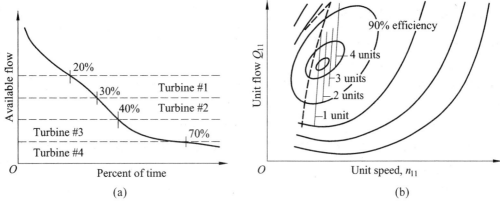

Fig.9.6 Effective use of multiple units

Questions

1. List factors affecting the selection of speed for a hydropower plant.

2. Define the term specific speed.

3. Compare the efficiencies of different types of hydraulic turbines, and provide their operating ranges.

4. Explain the general guideline of turbine selection.

5. Explain the criteria for selecting turbines in the overlapping zones.

References

[1] Edvard. What's important for water power designer to make the Right Choice of a Turbine [EB/OL]. (2013-10-25) [2016-06-16]. http://electrical-engineering-portal.com/whats-important-for-water-power-designer-to-make-the-right-choice-of-a-turbine.

[2] Mohammed Taih Catte and Rasim Azeez Kadhim. Chapter 4 Hydro Power [EB/OL]. (2012-10-31) [2016-06-16]. http://www.intechopen.com/books/energy-conservation/hydro-power.

[3] U.S. Department of the Interior. Selecting Hydraulic Reaction Turbines [M]. Denver, Colo.: U.S. Department of the Interior, Bureau of Reclamation, 1976.

[4] Indian Institute of Technology Roorkee, 3.1 Electro-Mechanical Selection of Turbine and Governing System [EB/OL]. (2012-06) [2016-06-16]. http://ahec.org.in/publ/standard/standard_pdf/3.1_Selection_of_turbine_and_governing_system.pdf.

[5] Walters R N, Bates C G. Selecting hydraulic reaction turbines[J]. NASA STI/Recon Technical Report N, 1976, 77: 20609.

[6] Warnick C C. Hydropower engineering[M]. New Jersey: Prenticc-Hall, Inc., Englewood Cliffs, 1984.

[7] Sangal S, Arpit G, Dinesh K. Review of Optimal Selection of Turbines for Hydroelectric Projects[J]. International Journal of Emerging Technology and Advance Engineering, 2013, 3: 424-430.

Important Words and Phrases

adjustable blade Kaplan turbine 轴流转桨式水轮机
alternative 比较方案
altitude 海拔；高度
capital cost 投资费用；资本成本
cavitation performance 空化性能
centerline 中心线
civil cost 土建成本
configuration 配置；结构
criterion 标准
critical value 临界值
English system 英制
field fabrication 现场制作
fixed blade propeller turbine 轴流定桨式水轮机
full gate opening 导叶全开度
generator rating 发电机额定值
geometry 几何形状
guideline 指导方针；指导原则
harness 利用
homologically 同源的

horsepower 马力
incur 招致，引发
instability 不稳定（性）
integrity 完整性
metric 公制的；米制的
model test 模型测试
offset 抵消；弥补
pitting 点状腐蚀；点蚀
plant factor 发电厂利用率；设备使用率
revolutions per minute 转每分钟
salient 显著的；突出的
setting depth 埋深；设置深度
setting 安装；布置
surge protection 过载保护
susceptibility 易受影响或损害的状态；敏感性
tailwater elevation 尾水位
Thoma cavitation coefficient 托马空化系数
throat 喉口
turbine setting 水轮机安装高程
vapor bubble 气泡

10 Classification and Components of Centrifugal Pumps

Centrifugal pumps are one of the most common components found in fluid systems. Centrifugal pumps enjoy widespread application partly due to their ability to operate over a wide range of flow rates and pump heads.

10.1 Classification by Flow

Centrifugal pumps can be classified based on the manner in which fluid flows through the pump. The manner in which fluid flows through the pump is determined by the design of the pump casing and the impeller. The three types of flow through a centrifugal pump are radial flow, axial flow, and mixed flow.

10.1.1 Radial Flow Pumps

In a radial flow pump, the liquid enters at the center of the impeller and is directed out along the impeller blades in a direction at right angles to the pump shaft. The impeller of a typical radial flow pump and the flow through a radial flow pump are shown in Fig.10.1.

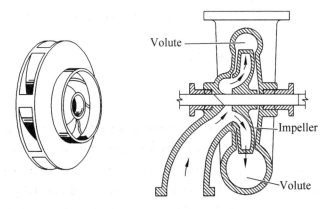

Fig.10.1　Radial flow centrifugal pump

10.1.2 Axial Flow Pumps

In an axial flow pump, the impeller pushes the liquid in a direction parallel to the pump shaft. Axial flow pumps are sometimes called propeller pumps because they operate essentially the same as the propeller of a boat. The impeller of a typical axial flow pump and the flow through a radial flow pump are shown in Fig.10.2.

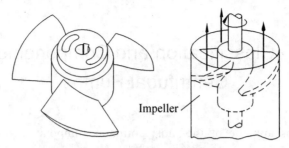

Fig.10.2 Axial flow centrifugal pump

10.1.3 Mixed Flow Pumps

Mixed flow pumps borrow characteristics from both radial flow and axial flow pumps. As liquid flows through the impeller of a mixed flow pump, the impeller blades push the liquid out away from the pump shaft and to the pump suction at an angle greater than 90°. The impeller of a typical mixed flow pump and the flow through a mixed flow pump are shown in Fig.10.3.

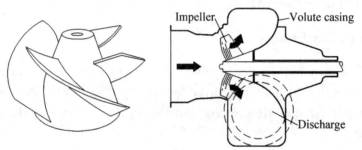

Fig.10.3 Mixed flow centrifugal pump

10.2 Stationary Components

10.2.1 Pump Casing and Volute

Centrifugal pumps basically consist of a stationary pump casing and an impeller mounted on a rotating shaft. The pump casing provides a pressure boundary for the pump and contains channels to properly direct the suction and discharge flow. The pump casing has suction and discharge penetrations for the main flow path of the pump and normally has small drain and vent fittings to remove gases trapped in the pump casing or to drain the pump casing for maintenance.

Fig.10.4 is a simplified diagram of a typical centrifugal pump that shows the relative locations of the pump suction, impeller, volute, and discharge.

The pump casing guides the liquid from the suction connection to the center, or eye, of the impeller. The vanes of the rotating impeller impart a radial and rotary motion to the liquid, forcing it to the outer periphery of the pump casing where it is collected in the outer part of the pump casing called the volute. The volute is a region that expands in cross-sectional area as it wraps around the pump casing. The purpose of the volute is to collect the liquid discharged from the periphery of the impeller at high velocity and gradually cause a reduction in fluid velocity by increasing the flow area. This converts the velocity head to static pressure. The fluid is then discharged from the pump through the discharge connection.

Centrifugal pumps can also be constructed in a manner that results in two distinct volutes, each receiving the liquid that is discharged from a 180° region of the impeller at any given time. Pumps of this type are called double volute

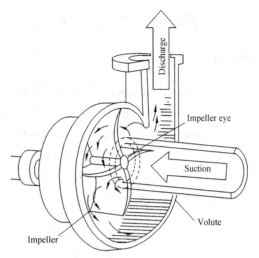

Fig.10.4　Centrifugal pump

pumps (they may also be referred to a split volute pumps). In some applications the double volute minimizes radial forces imparted to the shaft and bearings due to imbalances in the pressure around the impeller. A comparison of single and double volute centrifugal pumps is shown in Fig.10.5.

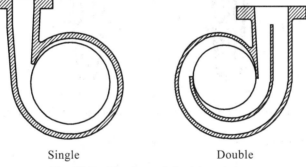

Fig.10.5　Single and double volutes

10.2.2　Diffuser

Some centrifugal pumps contain diffusers, as shown in Fig.10.6. A diffuser is a set of stationary vanes that surround the impeller. The purpose of the diffuser is to increase the efficiency of the centrifugal pump by allowing a more gradual expansion and less turbulent area for the liquid to reduce in velocity. The diffuser vanes are designed in a manner that the liquid exiting the impeller will encounter an ever-increasing flow area as it passes through the diffuser. This increase in flow area causes a reduction in flow velocity, converting kinetic energy into flow pressure.

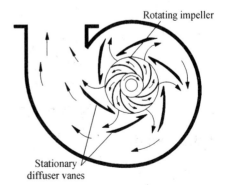

Fig.10.6　Centrifugal pump diffuser

10.3 Rotary Components

10.3.1 Impeller

Impellers of pumps are classified based on the number of points that the liquid can enter the impeller and also on the amount of webbing between the impeller blades.

Impellers can be either single-suction or double-suction. A single-suction impeller allows liquid to enter the center of the blades from only one direction. A double-suction impeller allows liquid to enter the center of the impeller blades from both sides simultaneously. Fig.10.7 shows simplified diagrams of single and double-suction impellers.

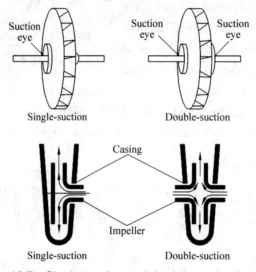

Fig.10.7 Single-suction and double-suction impellers

Impellers can be open, semi-open, or enclosed. The open impeller consists only of blades attached to a hub. The semi-open impeller is constructed with a circular plate (the web) attached to one side of the blades. The enclosed impeller has circular plates attached to both sides of the blades. Enclosed impellers are also referred to as shrouded impellers. Fig.10.8 illustrates examples of open, semi-open, and enclosed impellers.

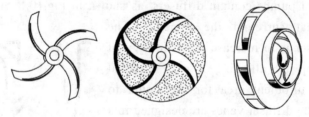

Fig.10.8 Open, semi-open, and enclosed impellers

10.3.2 Shaft

The function of a pump shaft is to transmit input power from the driver to the impeller mounted on the shaft. It is subjected to several stresses—flexural, shear, torsional, tensile, etc.

Of these, the torsional stress is most significant and is usually used as basis for sizing the shaft diameter. Commonly used shaft materials are 4140 carbon steel, and stainless steel such as 310, 410, or 416. The shaft position, horizontally or vertically, defines whether a pump is considered as a horizontal pump, or a vertical pump.

10.4 Shaft Seal Devices

Some of the most common shaft seal devices found in centrifugal pumps are stuffing boxes, packing, and lantern rings, as shown in Fig.10.9.

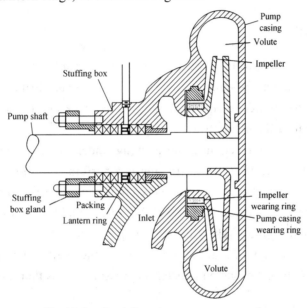

Fig.10.9 Centrifugal pump components

10.4.1 Stuffing Box

In almost all centrifugal pumps, the rotating shaft that drives the impeller penetrates the pressure boundary of the pump casing. It is important that the pump is designed properly to control the amount of liquid that leaks along the shaft at the point that the shaft penetrates the pump casing. There are many different methods of sealing the shaft penetration of the pump casing. Factors considered when choosing a method include the pressure and temperature of the fluid being pumped, the size of the pump, and the chemical and physical characteristics of the fluid being pumped.

One of the simplest types of shaft seal is the stuffing box. The stuffing box is a cylindrical space in the pump casing surrounding the shaft. Rings of packing material are placed in this space. Packing is material in the form of rings or strands that is placed in the stuffing box to form a seal to control the rate of leakage along the shaft. The packing rings are held in place by a gland. The gland is, in turn, held in place by studs with adjusting nuts. As the adjusting nuts are tightened, they move the gland in and compress the packing. This axial compression causes the packing to expand radially, forming a tight seal between the rotating shaft and the inside wall of the stuffing box.

The high speed rotation of the shaft generates a significant amount of heat as it rubs against the packing rings. If no lubrication and cooling are provided to the packing, the temperature of the packing increases to the point where damage occurs to the packing, the pump shaft, and possibly nearby pump bearings. Stuffing boxes are normally designed to allow a small amount of controlled leakage along the shaft to provide lubrication and cooling to the packing. The leakage rate can be adjusted by tightening and loosening the packing gland.

10.4.2 Lantern Ring

It is not always possible to use a standard stuffing box to seal the shaft of a centrifugal pump. The pump suction may be under a vacuum so that outward leakage is impossible or the fluid may be too hot to provide adequate cooling of the packing. These conditions require a modification to the standard stuffing box.

One method of adequately cooling the packing under these conditions is to include a lantern ring, as shown in Fig.10.9. A lantern ring is a perforated hollow ring located near the center of the packing box that receives relatively cool, clean liquid from either the discharge of the pump or from an external source and distributes the liquid uniformly around the shaft to provide lubrication and cooling. The fluid entering the lantern ring can cool the shaft and packing, lubricate the packing, or seal the joint between the shaft and packing against leakage of air into the pump in the event the pump suction pressure is less than that of the atmosphere.

10.4.3 Mechanical Seals

In some situations, packing material is not adequate for sealing the shaft. One common alternative method for sealing the shaft is using mechanical seal. Mechanical seal consists of two basic parts, a rotating element attached to the pump shaft and a stationary element attached to the pump casing. Each of these elements has a highly polished sealing surface. The polished faces of the rotating and stationary elements come into contact with each other to form a seal that prevents leakage along the shaft.

10.5 Wearing Rings

Centrifugal pumps contain rotating impellers within stationary pump casings. To allow the impeller to rotate freely within the pump casing, a small clearance is maintained between the impeller and the pump casing. To maximize the efficiency of a centrifugal pump, it is necessary to minimize the amount of liquid leaking through this clearance from the high pressure or discharge side of the pump back to the low pressure or suction side.

Some wear or erosion will occur at the point where the impeller and the pump casing nearly come into contact. This wear is due to the erosion caused by liquid leaking through this tight clearance and other causes. As wear occurs, the clearances become larger and the rate of leakage increases. Eventually, the leakage could become unacceptably large and maintenance would be required for the pump.

To minimize the cost of pump maintenance, many centrifugal pumps are designed with wearing rings, as shown in Fig.10.9. Wearing rings are replaceable rings that are attached to the impeller and/or the pump casing to allow a small running clearance between the impeller and the pump casing without causing wear of the actual impeller or pump casing material. These wearing rings are designed to be replaced periodically during the life of a pump and prevent the more costly replacement of the impeller or the casing.

10.6 Axial Thrust Balancing Devices

Axial thrust is a very important part of pump reliability. The very large surfaces of the impeller allow pressure to build up on the shrouds of the impeller causing very high loading of the bearing in the axial direction.

Different types of impellers have very different load characteristics. The open impeller with no shrouds, has very low axial thrust loading because there are no shrouds for the differential pressure to build upon. Semi-open impellers actually have the worst axial thrust characteristics because they have one shroud that allows the discharge pressure to build up across the whole shroud surface whereas on the other side of the shroud the pressure goes from the suction pressure and increases to the discharge pressure as water flows out radially, as shown in Fig.10.10. This differential pressure can often be in the thousands of pounds which would be directly applied on the bearings. Enclosed impellers with two shrouds make it much easier to balance the axial thrust but still require some form of a thrust balancing device.

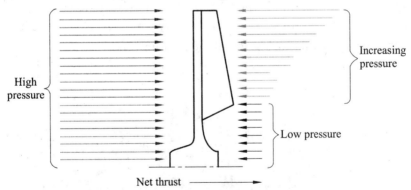

Fig.10.10 Axial thrust on a semi-open impeller

Axial thrust balancing devices include vanes on the back shroud of the impeller. These vanes pump the liquid from behind the impeller towards the outside diameter. This lowers the pressure near the shaft and increases the pressure towards the discharge diameter mimicking the thrust profile on the front of the impeller. This is often used for semi-open impeller designs.

Another popular form of axial thrust balancing device is to drill balance holes through the rear shroud of the impeller allowing the high pressure at the back of the impeller to bleed through the balance holes back into the suction, as shown in Fig.10.11. Most forms of thrust balancing cause a loss in efficiency, ultimately costing more money to operate the pump. They are necessary, however, for the mechanical integrity of the pump and the reliability of the unit.

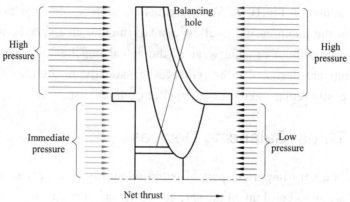

Fig.10.11 Axial thrust on an enclosed impeller

10.7 Multi-Stage Centrifugal Pumps

A centrifugal pump with a single impeller that can develop a differential pressure of more than 150 psid between the suction and the discharge is difficult and costly to design and construct. A more economical approach to developing high pressures with a single centrifugal pump is to include multiple impellers on a common shaft within the same pump casing. Internal channels in the pump casing route the discharge of one impeller to the suction of another impeller. Fig.10.12 shows a diagram of the arrangement of the impellers of a four-stage pump. The water enters the pump from the top left and passes through each of the four impellers in series, going from left to right. The water goes from the volute surrounding the discharge of one impeller to the suction of the next impeller.

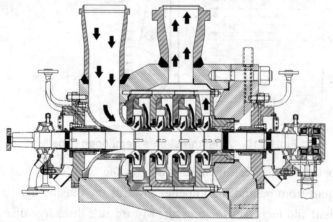

Fig.10.12 Multi-stage centrifugal pump

A pump stage is defined as that portion of a centrifugal pump consisting of one impeller and its associated components. Most centrifugal pumps are single-stage pumps, containing only one impeller. A pump containing seven impellers within a single casing would be referred to as a seven-stage pump, or generally as a multi-stage pump.

Questions

1. State the purposes of the following centrifugal pump components:
   - Impeller
   - Volute
   - Diffuser
   - Packing
   - Lantern ring
   - Wearing ring

2. Given a drawing of a centrifugal pump, identify the following major components:
   - Pump casing
   - Pump shaft
   - Impeller
   - Volute
   - Stuffing box
   - Stuffing box gland
   - Packing
   - Lantern ring
   - Impeller wearing ring
   - Pump casing wearing ring

References

[1] U.S. Department of Energy. DOE Fundamentals Handbook, Mechanical Science, Volume 1 of 2, Module 3 – Pumps: DOE-HDBK-1018/1-93 [S/OL]. Virginia: National Technical Information Service, U.S. Department of Commerce, 1993 [2016-06-16]. http://energy.gov/sites/prod/files/2013/06/f2/h1018v1.pdf.

[2] Centrifugal-Pump.ORG. Pump Basics [EB/OL]. [2016-06-16]. http://centrifugalpump.org/pump_basic.html.

[3] Intro To Pumps. Centrifugal Pump Fundamentals [EB/OL]. [2016-06-16]. http://www.introtopumps.com/pump-fundamentals/.

[4] Hydraulic Institute. About Rotodynamic Pumps [EB/OL]. [2016-06-16]. http://www.pumps.org/Pump_Fundamentals/Rotodynamic.aspx.

Important Words and Phrases

adjusting nut   调节螺母   axial flow pump   轴流泵

balance hole　平衡孔
balancing device　平衡装置
carbon steel　碳钢，碳素钢
centrifugal pump　离心泵
clearance　间隙；空隙
cooling　冷却
discharge connection　压水管
discharge flow　排出流
double volute pump　双蜗壳泵
double-suction impeller　双吸式叶轮
double-suction　双吸
drain fitting　排水配件
enclosed impeller　封闭式叶轮
erosion　侵蚀，腐蚀
ever-increasing　不断增加的
flexural stress　挠曲应力
flow area　过水面积；流动面积
gland　压盖
head　扬程
horizontal pump　卧式泵
impeller eye　叶轮入口
in series　串联
lantern ring　水封环；套环
leak　使渗漏，泄露
mechanical seal　机械密封
multi-stage centrifugal pump　多级离心泵
open impeller　敞开式叶轮
packing material　填充材料
packing ring　填料环；密封圈
penetrate　穿透
propeller pump　旋桨泵；轴流泵
pump casing　泵壳

pump shaft　泵轴
radial flow pump　径流泵
route　按某路线发送；给……规定路线
rub　摩擦
semi-open impeller　半开式叶轮
shaft seal device　轴封装置
shaft seal　轴封，轴封装置；主轴密封
shear stress　剪切应力
shroud　盖板
shrouded impeller　闭式叶轮
single-stage pump　单级泵
single-suction impeller　单吸式叶轮
single-suction　单吸
split volute pump　剖分式蜗壳泵
static pressure　静压
stationary element　固定元件；固定部分
strand　股（绳子的）
stress　应力；压力
stud　螺柱
suction connection　吸水管
suction flow　吸入流
tensile stress　拉应力
torsional stress　扭曲应力
turbulent　湍流的
vent fitting　排气装置
vertical pump　立式泵
volute casing　蜗形机壳；蜗壳
volute　蜗壳；螺旋形；涡形
wear　磨损；耗损
wearing ring　承磨环；减磨环，抗磨环
web　盖板；腹板
webbing　盖板；边带

11 Basic Parameters of Centrifugal Pumps

To understand the operating principles of centrifugal pumps, we must first understand the basic performance parameters used in centrifugal pumps.

11.1 Flow Rate

11.1.1 Volumetric Flow Rate

The first and most important point to consider is that centrifugal pumps are volumetric machines. The flow rate of a pump is the amount of liquid conveyed per unit time. It is actually the volumetric rate of flow. Other common terms for flow rate are capacity and discharge rate. The classical English unit is gallons per minute (gpm). The metric equivalents are liters per minute (L/min) or cubic meters per second (m³/sec) or cubic meters per hour (m³/h). The flow rate will be denoted as Q.

It is worthy to note that any pump, for a single point of operation would always give the same volumetric flow rate for any liquid, be it hydrocarbon, water or any other. Depending on the density of the fluid, the mass flow rate changes.

11.1.2 Mass Flow Rate

The mass flow rate q is the usable mass of liquid per unit time discharged by the pump through its outlet branch. The mass flow rate changes with the density of the fluid. The common units are kg/s and t/h.

The relationship between mass flow rate q and volumetric flow rate Q is

$$q = \rho Q \quad (\rho = \text{density of liquid})$$

11.2 Head

Head is simply a pressure unit that is commonly used in hydraulic engineering. It is a term used to express pressure in both pump design and system design when analyzing static or dynamic conditions. It will be denoted as H. Its unit is feet (meters in the metric system of units).

This relationship is expressed as:

$$\text{head in feet} = \frac{(\text{pressure in psi} \times 2.31)}{\text{specific gravity}}$$

where specific gravity is the ratio of the density of a liquid over the density of water. Water is designated with a specific gravity of 1.0.

Pressure in static systems is referred to as static head and in dynamic systems as dynamic head.

11.2.1 System Head

All forms of energy involved in a liquid flow system can be expressed in terms of feet of liquid. The total of these various heads determines the total system head or the work which a pump must perform in the system. The various forms of head are defined as follows.

(1) Total static head is the vertical distance between the surface of the suction source liquid and the surface level of the discharge liquid.

(2) Static discharge head is the vertical distance from the centerline of the suction nozzle up to the surface level of the discharge liquid.

(3) Static suction head applies when the supply is above the pump. It is the vertical distance from the centerline of the suction nozzle up to the liquid surface of the suction supply.

(4) Static suction lift applies when the supply is located below the pump. It is the vertical distance from the centerline of the suction nozzle down to the surface of the suction supply liquid.

The above four terms are illustrated in Fig.11.1.

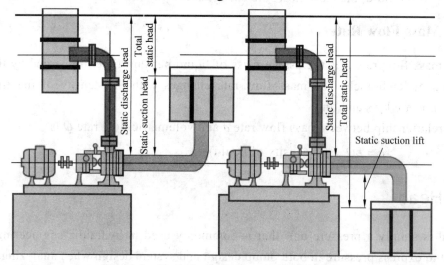

Fig.11.1 Static heads of a pumping system

Velocity, friction, and pressure head are used in conjunction with static heads to define dynamic heads.

(5) Velocity head (h_v) is the energy in a liquid as a result of it traveling at some velocity v. It can be thought of as the vertical distance a liquid would need to fall to gain the same velocity as a liquid traveling in a pipe. This relationship is expressed as:

$$h_v = v^2/2g$$

where

    h_v—velocity head;
    v—velocity of the liquid in feet per second;
    g—32.2 ft/sec^2.

The velocity head is usually insignificant and can be ignored in most high head systems. However, it can be a large factor and must be considered in low head systems.

(6) Friction head (h_f) is the head needed to overcome resistance to liquid flowing in a system. It is dependent upon the size, condition and type of pipe, number and type of pipe fittings, flow rate, and nature of the liquid.

(7) Pressure head must be considered when a pumping system either begins or terminates in a tank which is under some pressure other than atmospheric. A vacuum in the suction tank or a positive pressure in the discharge tank must be added to the system head, whereas a positive pressure in the suction tank or vacuum in the discharge tank would be subtracted. If a vacuum exists and the value is known in inches of mercury, the equivalent feet of liquid can be calculated using the following formula:

$$vacuum\ in\ feet = \frac{in.\ of\ Hg \times 1.13}{specific\ gravity}$$

It is important to convert this pressure into feet of liquid when analyzing systems so that all units are the same.

The above forms of head, namely static, friction, velocity, and pressure, are combined to make up the total system head at any particular flow rate.

11.2.2 Pump Head

When discussing how a pump performs in service, we use terms describing dynamic head. In other words, when a pump is running, it is dynamic. Pumping systems are also dynamic when liquid is flowing through them, and they must be analyzed as such. To do this, the following four dynamic terms are used.

(1) Total dynamic suction head (h_s) is the static suction head plus the velocity head at the suction flange minus the total friction head in the suction line. The total dynamic suction head, as determined on pump test, is the reading of the gauge on the suction flange, converted to feet of liquid and corrected to the pump centerline, plus the velocity head at the point of gauge attachment.

(2) Total dynamic discharge head (h_d) is the static discharge head plus the velocity head at the pump discharge flange plus the total friction head in the discharge system. The total dynamic discharge head, as determined on pump test, is the reading of a gauge at the discharge flange, converted to feet of liquid and corrected to the pump centerline, plus the velocity head at the point of gauge attachment.

(3) Total dynamic suction lift (h_s) is the static suction lift minus the velocity head at the

suction flange plus the total friction head in the suction line. To calculate the total dynamic suction lift, take suction pressure at the pump suction flange (the reading of a gauge on the suction flange), convert it to head and correct it to the pump centerline, then subtract the velocity head at the point of gauge attachment.

(4) Total dynamic head (*TDH*) in a system is the total dynamic discharge head minus the total dynamic suction head when the suction supply is above the pump.

$$TDH = h_d - h_s \quad \text{(with a suction head)}$$

When the suction supply is below the pump, the total dynamic head is the total dynamic discharge head plus the total dynamic suction lift.

$$TDH = h_d + h_s \quad \text{(with a suction lift)}$$

11.2.3 Prediction of *TDH*

Centrifugal pumps are dynamic machines that impart energy to liquids. This energy is imparted by changing the velocity of the liquid as it passes through the impeller. Most of this velocity energy is then converted into pressure energy (total dynamic head) as the liquid passes through the casing or diffuser.

To predict the approximate total dynamic head of any centrifugal pump, we must go through two steps.

First, the velocity at the outside diameter of the impeller is calculated using the following formula:

$$v = \frac{RPM \times D}{229}$$

where
  v—velocity at the periphery of the impeller in ft/sec;
  D—outside diameter of the impeller in inches;
  RPM—revolutions per minute of the impeller;
  229—a constant.

Second, because the velocity energy at the outside diameter or periphery of the impeller is approximately equal to the total dynamic head developed by the pump, we continue by substituting v from above into the following equation:

$$H = v^2/2g$$

where
  H—total dynamic head developed in ft;
  v—velocity at the outside diameter of the impeller in ft/sec;
  g—32.2 ft/sec^2.

A centrifugal pump operating at a given speed and impeller diameter will raise liquid of any specific gravity or weight to a given height. Therefore, we always think in terms of feet of liquid rather than pressure when analyzing centrifugal pumps and their systems.

11.3 Power

In physics, power is defined as work per unit time. In the field of engineering, power is defined as the ability to do work. Units for power are the horsepower (hp) and the kilowatt (kW).

When we discuss horsepower there exists no less than three different horsepowers involved in centrifugal pump systems. These are hydraulic horsepower, brake horsepower, and drive or motor horsepower.

11.3.1 Hydraulic Horsepower

Pump output or hydraulic horsepower or water horsepower (*WHP*) is the liquid horsepower delivered by the pump to the liquid. It is defined by the following formula,

$$WHP = \frac{Q \times TDH \times specific\ gravity}{3960} \text{ in hp}\quad \text{or} \quad WHP = \frac{Q \times TDH \times \rho}{367} \text{ in kW}$$

where
  Q—flow rate, gpm or m³/h;
  H—head, feet of liquid or m;
  ρ—density of liquid, kg/dm³;
  3,960—constant, obtained by dividing the number or foot pounds for one horsepower (33,000) by the weight of one gallon of water (8.33 pounds).

11.3.2 Brake Horsepower

To provide a certain amount of power to the liquid, a larger amount of power must be provided to the pump shaft to overcome inherent losses. Pump input or brake horsepower (*BHP*) is the actual horsepower delivered to the pump shaft. It is higher than the hydraulic horsepower by the amount of the pump losses. It is defined by the following formula,

$$BHP = \frac{Q \times TDH \times specific\ gravity}{3\,960 \times pump\ efficiency}$$

Brake horsepower is the quantity that is generally provided by pump manufacturers on performance curves. For this reason, horsepower or simply power, shall be taken to mean brake horsepower in this book.

11.3.3 Drive horsepower

Prime movers, also known as drives, are machines that convert natural energy into work.

Drive horsepower is the nominal or nameplate power rating of the prime mover. The two primary types of drives for centrifugal pumps are electric motors and steam turbines.

11.4 Efficiency

The overall efficiency of a centrifugal pump is the ratio of the power input over the pump shaft to that of the power transferred to the liquid, and it will be denoted with the symbol η.

$$\eta = \frac{hydraulic\ horsepower}{brake\ horsepower} = \frac{WHP}{BHP} = \frac{Q \times TDH \times specific\ gravity}{3\ 960 \times BHP}$$

The pump does not completely convert kinetic energy to pressure energy since some of the kinetic energy is lost in this process. Primarily, there are three areas where this energy is dissipated and not converted to useful work. Pump efficiency is a factor that accounts for all these losses which will be discussed in the chapter of Pump Losses.

11.4.1 Hydraulic Efficiency

Hydraulic efficiency is used to measure the consumption of power by disk friction and shock losses which values are calculated in the chapter of Characteristics of Centrifugal Pumps and Their Systems. Disk friction is the main cause of power consumption, which is the friction of the liquid with the impeller shrouds. This is a function of speed and impeller geometry. Shock losses refer to the losses caused by rapid changes in direction along the impeller and volute.

$$\eta_h = \frac{TDH}{H_{theo}}$$

where

η_h —hydraulic efficiency;

H_{theo} —the theoretic head of a pump without considering any losses.

11.4.2 Volumetric Efficiency

Volumetric efficiency is used to measure the recirculation losses at wear rings, interstage bushes, balancing holes and vane clearances in the case of semi-open impellers. It refers to the percentage of actual fluid flow out of the pump over the flow out of the pump without leakage.

$$\eta_v = \frac{Q}{Q_{theo}}$$

where

η_v —volumetric efficiency;

Q_{theo} —the theoretic flow rate of a pump without considering any leakages;

Q—the actual flow rate of a pump.

11.4.3 Mechanical Efficiency

Mechanical efficiency is used to measure the mechanical losses due to friction at seals or gland packing and bearings.

$$\eta_m = \frac{N_{theo}}{BHP}$$

where

η_m —mechanical efficiency;

N_{theo} — total theoretic power transferred to the liquid by the impeller,

$$N_{theo} = \frac{Q_{theo} \times H_{theo} \times specific\ gravity}{3\ 960}$$

The total efficiency is a product of the above three efficiencies:

$$\eta = \eta_h \times \eta_v \times \eta_m$$
$$= \frac{TDH}{H_{theo}} \times \frac{Q}{Q_{theo}} \times \frac{Q_{theo} \times H_{theo} \times specific\ gravity}{3\ 960 \times BHP}$$
$$= \frac{Q \times TDH \times specific\ gravity}{3\ 960 \times BHP}$$

Although mechanical and volumetric losses are important components, hydraulic efficiency is the largest factor.

11.5 Rotational Speed

Rotational speed is generally referred to simply as speed. The unit of revolutions per minute (rpm) is used in conjunction with speed. Speed is denoted by the symbol n.

11.6 Net Positive Suction Head (*NPSH*)

Net Positive Suction Head or *NPSH* for pumps can be defined as the difference between liquid pressure at pump suction and liquid vapor pressure, expressed in terms of height of liquid column. It will be discussed in detail in the chapter of Cavitation and *NPSH* in Centrifugal Pumps.

Questions

1. List the basic parameters of centrifugal pumps.
2. Define the following terms:
 · Total static head
 · Static discharge head

· Static suction head
· Static suction lift
· Hydraulic horsepower
· Brake horsepower
· *NPSH*

3. Calculate and predict the total dynamic head of a pumping system.

4. Give examples for losses affecting the hydraulic efficiency, volumetric efficiency and mechanical efficiency of pumps.

5. Derive the equation of total efficiency for a centrifugal pump by hydraulic efficiency, volumetric efficiency and mechanical efficiency.

References

[1] Bachus, Larry, and Angel Custodio. Know and understand centrifugal pumps[M]. Amsterdam: Elsevier, 2003.

[2] Girdhar P, Moniz O. Practical centrifugal pumps[M]. Amsterdam: Elsevier, 2011.

[3] Sterling SIHI. Basic principles for the design of centrifugal pump installations[M]. Zoetermeer: Sterling Fluid Systems Group, 2003.

[4] Magdy Abou Rayan, Nabil H. Mostafa, Purage Ohans. Textbook of Machines Hydraulic [M]. Zagazig: Zagazig University, Dept. of Mechanical Engineering, 2009.

[5] GRUNDFOS Management A/S. The Centrifugal Pump[M]. Bjerringbro: GRUNDFOS Research and Technology, 2009.

[6] Joe Rvans. Centrifugal Pump Efficiency-What Is Efficiency [EB/OL]. [2016-06-16]. http://www.pumpsandsystems.com/topics/pumps/pumps/centrifugal-pump-efficiency-what-efficiency.

Important Words and Phrases

angular velocity　角速度
brake horsepower　制动功率
denote　表示
designate　指定
discharge rate　流量
discharge tank　压水池
disk friction　圆盘摩擦
dissipate　消散
drive horsepower　驱动功率
drive　驱动器
dynamic head　动水头
electric motor　电动机
English unit　英制单位

feet　英尺
friction head　摩擦头；摩擦水头
gain　获得；增加
gallons per minute　加仑每分钟
gauge　计量器
gland packing　压盖填料
gpm　加仑每分钟
hydraulic horsepower　水力功率
hydrocarbon　碳氢化合物
inch　英寸
inherent loss　固有损耗
interstage bush　级间衬套
mass flow rate　质量流量

mechanical efficiency 机械效率
mechanical loss 机械损失
mercury 汞，水银
minus 减去
motor horsepower 电机功率
nameplate 铭牌
Net Positive Suction Head 气蚀余量
nominal 名义上的
NPSH 气蚀余量
outlet branch 出水支管
performance curve 性能曲线
power rating 额定功率
power 功率
prime mover 原动机
product 乘积
pump input 水泵输入功率
pump output 水泵输出功率
recirculation loss 回流损失
scalar quantity 标量
shock loss 冲击损失
specific gravity 比重
static discharge head 静排出压头，压水地形高度
static head 静水头
static suction head 静吸入压头；吸水地形高度
static suction lift 静吸上高度；静止吸入高度
substitute 代替；替换
subtract 减去
suction flange 进口法兰
suction line 吸水管线
suction nozzle 吸水口
suction tank 吸水池
terminate 结束，终止
total dynamic discharge head 总动排出压头
total dynamic head 总动压头
total dynamic suction head 总动吸入压头
total dynamic suction lift 总动吸上高度；总动吸入高度
total static head 总静水头
vapor pressure 蒸汽压
volumetric efficiency 容积效率
volumetric flow rate 体积流量
volumetric 体积的；容积的
water horsepower 水功率

12 Centrifugal Pump Theory

This chapter describes the theoretical foundation of energy conversion in a centrifugal pump. When the pump operates, energy is added to the shaft in the form of mechanical energy. In the impeller it is converted to internal (static pressure) and kinetic energy (velocity). The process is described through Euler's pump equation. By means of velocity triangles for the flow in the impeller inlet and outlet, the pump equation can be interpreted and a theoretical loss-free head and power consumption can be calculated.

Velocity triangles can also be used for prediction of the pump's performance in connection with changes of e.g. speed, impeller diameter and width.

12.1 Velocity Triangles

An example of velocity triangles is shown in Fig.12.1. Here U describes the impeller's tangential velocity while the absolute velocity C is the fluid's velocity compared to the surroundings. The relative velocity W is the fluid velocity compared to the rotating impeller. The angles α and β describe the fluid's absolute and relative flow angles respectively compared to the tangential direction. These velocity vectors are added through vector addition, forming velocity triangles at the inlet and outlet of the impeller.

$$\vec{C} = \vec{U} + \vec{W}$$

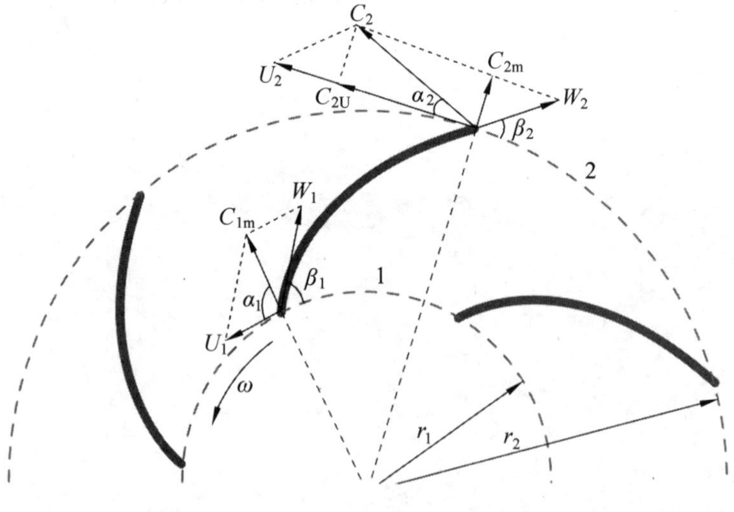

Fig.12.1 Velocity triangles positioned at the impeller inlet and outlet

By drawing the velocity triangles at inlet and outlet, the performance curves of the pump can be calculated by means of Euler's pump equation.

12.1.1 Inlet

Usually it is assumed that the flow at the impeller inlet is non-rotational. This means that $\alpha_1 = 90°$. The triangle is drawn as shown in Fig.12.1 position 1, and C_{1m} is calculated from the flow and the ring area in the inlet.

The ring area can be calculated in different ways depending on impeller type (radial impeller or semi-axial impeller), see Fig.12.2. For a radial impeller this is:

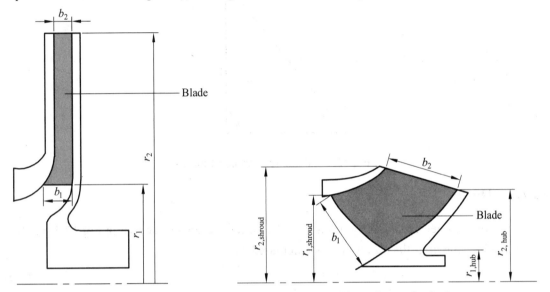

Fig.12.2 Radial impeller at the top, semi-axial impeller at the bottom

$$A_1 = 2\pi \cdot r_1 \cdot b_1 \quad (\text{m}^2) \tag{12.1}$$

where
r_1—The radial position of the impeller's inlet edge (m)
b_1—The blade's height at the inlet (m)

and for a semi-axial impeller this is:

$$A_1 = 2\pi \cdot \left(\frac{r_{1,\text{hub}} + r_{1,\text{shroud}}}{2} \right) \cdot b_1 \quad (\text{m}^2) \tag{12.2}$$

The entire flow must pass through this ring area. C_{1m} is then calculated from:

$$C_{1m} = \frac{Q_{\text{impeller}}}{A_1} \quad (\text{m/s}) \tag{12.3}$$

The tangential velocity U_1 equals the product of radius and angular frequency:

$$U_1 = 2\pi \cdot r_1 \cdot \frac{n}{60} = r_1 \cdot \omega \quad (\text{m/s}) \tag{12.4}$$

where
ω—Angular frequency (s^{-1});

n — Rotational speed (min⁻¹).

When the velocity triangle has been drawn, see Fig.12.3, based on α_1, C_{1m} and U_1, the relative flow angle β_1 can be calculated. Without inlet rotation ($C_1 = C_{1m}$) this becomes:

$$\tan \beta_1 = \frac{C_{1m}}{U_1} \tag{12.5}$$

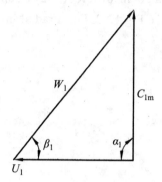

Fig.12.3 Velocity triangle at inlet

12.1.2 Outlet

As with the inlet, the velocity triangle at the outlet is drawn as shown in Fig.12.1 position 2. For a radial impeller, the outlet area is calculated as:

$$A_2 = 2\pi \cdot r_2 \cdot b_2 \quad (\text{m}^2) \tag{12.6}$$

and for a semi-axial impeller this is:

$$A_2 = 2\pi \cdot \left(\frac{r_{2,\text{hub}} + r_{2,\text{shroud}}}{2}\right) \cdot b_2 \quad (\text{m}^2) \tag{12.7}$$

C_{2m} is then calculated in the same way as for the inlet:

$$C_{2m} = \frac{Q_{\text{impeller}}}{A_2} \quad (\text{m/s}) \tag{12.8}$$

The tangential velocity U_2 is calculated from the following:

$$U_2 = 2\pi \cdot r_2 \cdot \frac{n}{60} = r_2 \cdot \omega \quad (\text{m/s}) \tag{12.9}$$

In the beginning of the design phase, β_2 is assumed to have the same value as the blade angle. The relative velocity can then be calculated from:

$$W_2 = \frac{C_{2m}}{\sin \beta_2} \quad (\text{m/s}) \tag{12.10}$$

and C_{2U} as:

$$C_{2U} = U_2 - \frac{C_{2m}}{\tan \beta_2} \quad (\text{m/s}) \tag{12.11}$$

Here by the velocity triangle at the outlet has been determined and can now be drawn, see Fig.12.4.

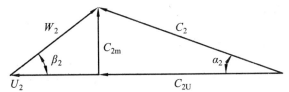

Fig.12.4 Velocity triangle at outlet

12.2 Euler's Pump Equation

Euler's pump equation is the most important equation in connection with pump design. The equation can be derived in many different ways. The method described here includes a control volume which limits the impeller, the moment of momentum equation which describes flow forces and velocity triangles at inlet and outlet.

A control volume is an imaginary limited volume which is used for setting up equilibrium equations. The equilibrium equations can be set up for torques, energy and other flow quantities which are of interest. The moment of momentum equation is one such equilibrium equation, linking mass flow and velocities with the impeller diameter. A control volume between 1 and 2, as shown in Fig.12.5, is often used for an impeller.

The balance which we are interested in is a torque balance. The torque (T) from the drive shaft corresponds to the torque originating from the fluid's flow through the impeller with mass flow $m=rQ$:

$$T = m \cdot (r_2 \cdot C_{2U} - r_1 \cdot C_{1U}) \quad (N \cdot m) \quad (12.12)$$

By multiplying the torque by the angular velocity, an expression for the shaft power (P_2) is found. At the same time, radius multiplied by the angular velocity equals the tangential velocity, $r_2\omega = U_2$. This results in:

$$\begin{aligned}
P_2 &= T \cdot \omega \quad (W) \quad (12.13)\\
&= m \cdot \omega \cdot (r_2 \cdot C_{2U} - r_1 \cdot C_{1U})\\
&= m \cdot (\omega \cdot r_2 \cdot C_{2U} - \omega \cdot r_1 \cdot C_{1U})\\
&= m \cdot (U_2 \cdot C_{2U} - U_1 \cdot C_{1U})\\
&= Q \cdot \rho \cdot (U_2 \cdot C_{2U} - U_1 \cdot C_{1U})
\end{aligned}$$

According to the energy equation, the hydraulic power added to the fluid can be written as the increase in pressure ΔP_{tot} across the impeller multiplied by the flow Q:

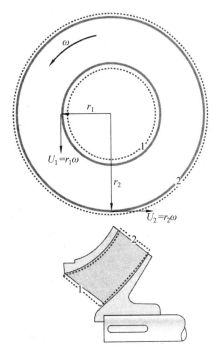

Fig.12.5 Control volume for an impeller

$$P_{hyd} = \Delta P_{tot} \cdot Q \quad (W) \tag{12.14}$$

The head is defined as:

$$H = \frac{\Delta P_{tot}}{\rho \cdot g} \quad (m) \tag{12.15}$$

and the expression for hydraulic power can therefore be transcribed to:

$$P_{hyd} = Q \cdot H \cdot \rho \cdot g = m \cdot H \cdot g \quad (W) \tag{12.16}$$

If the flow is assumed to be loss free, then the hydraulic and mechanical power can be equated:

$$P_{hyd} = P_2$$
$$m \cdot H \cdot g = m \cdot (U_2 \cdot C_{2U} - U_1 \cdot C_{1U})$$
$$H = \frac{(U_2 \cdot C_{2U} - U_1 \cdot C_{1U})}{g} \tag{12.17}$$

This is the equation known as Euler's equation, and it expresses the impeller's head at tangential and absolute velocities in inlet and outlet.

If the cosine relations are applied to the velocity triangles, Euler's pump equation can be written as the sum of the three contributions:

- $\dfrac{U_2^2 - U_1^2}{2g}$ —Static head as consequence of the centrifugal force;

- $\dfrac{W_2^2 - W_1^2}{2g}$ —Static head as consequence of the velocity change through the impeller;

- $\dfrac{C_2^2 - C_1^2}{2g}$ —Dynamic head.

$$H = \frac{U_2^2 - U_1^2}{2g} + \frac{W_2^2 - W_1^2}{2g} + \frac{C_2^2 - C_1^2}{2g} \quad (m) \tag{12.18}$$

If there is no flow through the impeller and it is assumed that there is no inlet rotation, then the head is only determined by the tangential velocity based on Eq.12.17 where $C_{2U} = U_2$:

$$H_0 = \frac{U_2^2}{g} \quad (m) \tag{12.19}$$

When designing a pump, it is often assumed that there is no inlet rotation meaning that C_{1U} equals zero.

$$H = \frac{U_2 \cdot C_{2U}}{g} \quad (m) \tag{12.20}$$

12.3 Blade Shape and Pump Curve

If it is assumed that there is no inlet rotation ($C_{1U} = 0$), a combination of Euler's pump

Eq.12.17 and Eq.12.6, 12.8 and 12.11 show that the head varies linearly with the flow, and that the slope depends on the outlet angle β_2:

$$H = \frac{U_2^2}{g} - \frac{U_2}{\pi \cdot D_2 \cdot b_2 \cdot g \cdot \tan(\beta_2)} \cdot Q \quad \text{(m)} \tag{12.21}$$

Fig.12.6 and Fig.12.7 illustrate the connection between the theoretical pump curve and the blade shape indicated at β_2.

Real pump curves are, however, curved due to different losses, slip, inlet rotation, etc.

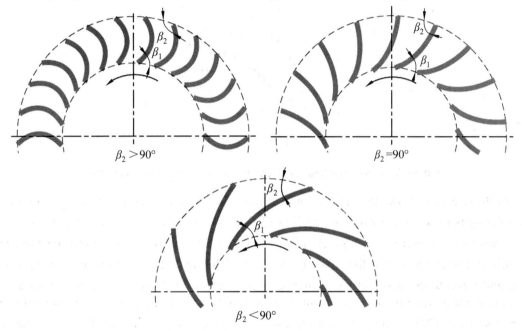

Fig.12.6 Blade shapes depending on outlet angle

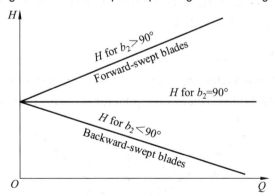

Fig.12.7 Theoretical pump curves

12.4 Slip

In the derivation of Euler's pump equation it is assumed that the flow follows the blade. In reality this is, however, not the case because the flow angle usually is smaller than the

111

blade angle. This condition is called slip. Nevertheless, there is a close connection between the flow angle and the blade angle. An impeller has an endless number of blades which are extremely thin, then the flow lines will have the same shape as the blades, see Fig.12.8.

The flow will not follow the shape of the blades completely in a real impeller with a limited number of blades with finite thickness. The tangential velocity out of the impeller as well as the head is reduced due to this. One possible explanation for slip is given as follows.

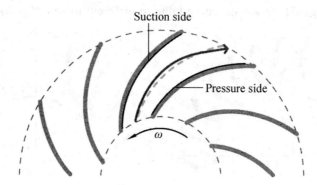

Fig.12.8 Ideal flow line: dashed line, actual flow line: solid line

In the course of flow through the impeller passage, there occurs a difference in pressure and velocity between the leading and trailing faces of the impeller blades. On the leading face of a blade there is relatively a high pressure and low velocity, while on the trailing face, the pressure is lower and hence the velocity is higher. This results in a circulation around the blade and a non-uniform velocity distribution at any radius. The mean direction of flow at the outlet, under this situation, changes from the blade angle at outlet β_2 to a different angle β_2' as shown in Fig.12.9. Therefore the tangential velocity component at outlet C_{2U} is reduced to C_{2U}', as shown by the velocity triangles in Fig.12.9, and the difference ΔC_{2U} is defined as the slip. The slip factor σ_S is defined as

$$\sigma_S = C_{2U}' / C_{2U} \tag{12.22}$$

With the application of slip factor σ_S, the work head imparted to the fluid (Euler head) becomes $\sigma_S C_{2U} U_2 / g$. The typical values of slip factor lie in the region of 0.9.

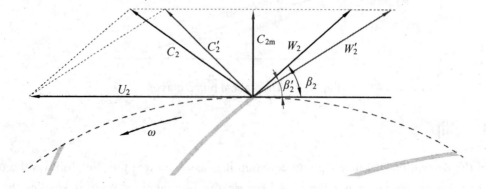

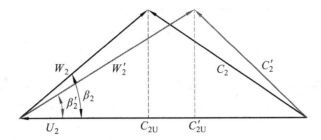

Fig.12.9 Velocity triangles where ' indicates the velocity with slip

12.5 Specific Speed of Centrifugal Pumps

The concept of specific speed for a pump is the same as that for a turbine. However, the quantities of interest are n_s, H and Q rather than n, H and P like in case of a turbine.

For pump

$$n_s = nQ^{1/2} / H^{3/4} \tag{12.23}$$

where

n_s—specific speed (unitless);

n—pump rotational speed (radians per second);

Q—flow rate (m³/s) at the point of best efficiency;

H—total head (m) per stage at the point of best efficiency.

The effect of the shape of rotor on specific speed is also similar to that for turbines. That is, radial flow (centrifugal) impellers have the lower values of n_s compared to those of axial-flow designs. The impeller, however, not the entire pump and, in particular, the shape of volute may appreciably affect the specific speed. Nevertheless, in general, centrifugal pumps are best suited for providing high heads at moderate rates of flow as compared to axial flow pumps which are suitable for large rates of flow at low heads. Similar to turbines, the higher is the specific speed, the more compact is the machine for given requirements. For multistage pumps, the specific speed refers to a single stage.

The impeller and the shape of the pump curves can be predicted based on the specific speed, see Fig.12.10.

Pumps with low specific speed, so-called low n_s pumps, have a radial outlet with large outlet diameter compared to inlet diameter. The head curves are relatively flat, and the power curve has a positive slope in the entire flow area.

On the contrary, pumps with high specific speed, so-called high n_s pumps, have an increasingly axial outlet, with small outlet diameter compared to the width. Head curves are typically descending and have a tendency to create saddle points. Performance curves decrease when flow increases. Different pump sizes and pump types have different maximum efficiencies.

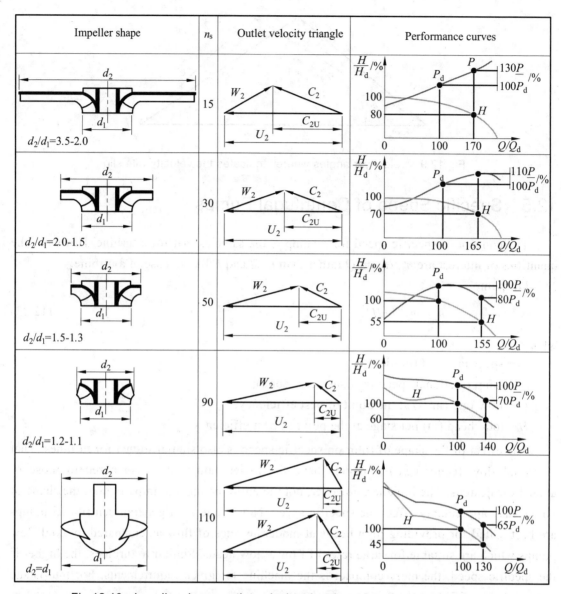

Fig.12.10 Impeller shape, outlet velocity triangle and performance curve as function of specific speed n_s

Questions

1. Draw the velocity triangles at the impeller inlet and outlet, and explain different velocity components.

2. Derive the Euler's equation.

3. Illustrate the relationship between the blade shape and the pump curve.

4. Define the term slip.

5. Define the term specific speed and illustrate the influences of specific speed on the impeller shape and the performance curve.

References

[1] NPTEL. Module 6 Pumps [EB/OL]. [2016-06-16]. http://nptel.ac.in/courses/Webcourse-contents/IIT-KANPUR/machine/ui/Course_home-8.htm.

[2] Codecogs. Centrifugal Pumps [EB/OL]. (2011-10-09) [2016-06-16]. http://www.codecogs.com/library/engineering/fluid_mechanics/machines/pumps/centrifugal-pumps.php.

[3] GRUNDFOS Management A/S. The Centrifugal Pump[M]. Bjerringbro: GRUNDFOS Research and Technology, 2009.

Important Words and Phrases

absolute velocity 绝对速度
angular frequency 角频率
blade angle 叶片安放角
blade shape 叶片形状
control volume 控制体积；控制体
cosine 余弦
dashed line 虚线
derive 推导出；得到，导出
descend 下降
dynamic head 动扬程
energy conversion 能量转换
energy equation 能量方程
equilibrium equation 平衡方程
Euler 欧拉
Euler's equation 欧拉方程
expression 表达式
finite 有限的
flow angle 进水角
head curve 扬程曲线
impeller passage 叶轮流道；叶槽
inlet rotation 入口旋转；入口回旋；进口旋转
interpret 解释
leading face 迎水面；工作面；正面
loss 损失
mass flow 质量流量

moment of momentum equation 动量矩方程
momentum equation 动量方程
multistage pump 多级泵
non-rotational 无旋的；非旋转的
power consumption 功率消耗
power curve 功率曲线
pressure side 迎水面；叶片正面；叶片工作面
real pump curve 泵的实际曲线
relative velocity 相对速度
ring area 环形区域
saddle point 鞍点
slip factor 反旋系数
slip 反旋
solid line 实线
static head 静扬程
suction side 背水面；叶片背面；叶片负压侧
tangential velocity 切线速度；切向速度
theoretical foundation 理论基础
theoretical loss-free head 无损失理论扬程
theoretical pump curve 泵的理论曲线
trailing face 背水面
vector addition 向量加法；矢量加法
velocity triangle 速度三角形
velocity vector 速度矢量

13 Pump Losses

Euler's pump equation provides a simple, loss-free description of the impeller performance. In reality, because of a number of mechanical and hydraulic losses in impeller and pump casing, the pump performance is lower than that predicted by the Euler's pump equation. The losses cause smaller head than the theoretical and higher power consumption, see Figs.13.1 and 13.2. The result is a reduction in efficiency.

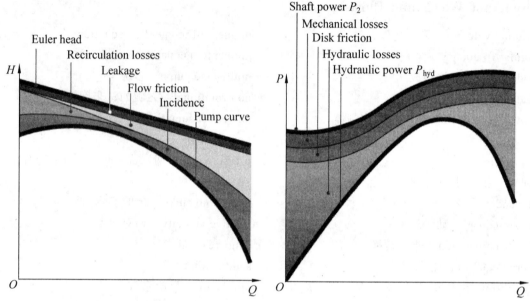

Fig.13.1   Reduction of theoretical Euler head due to losses

Fig.13.2   Increase in power consumption due to losses

13.1 Loss Types

Distinction is made between two primary types of losses: mechanical losses and hydraulic losses which can be divided into a number of subgroups. Mechanical losses can be further divided into bearing loss and shaft seal loss which will cause high power consumption. Hydraulic losses can be further classified into flow friction loss, mixing loss, recirculation loss, incidence loss, disk friction loss and leakage loss. The former four kinds of hydraulic losses will lower the head, the disk friction loss will cause high power consumption, and the leakage loss will reduce the flow rate.

Pump performance curves can be predicted by means of theoretical or empirical calculation models for each single type of loss. Accordance with the actual performance curves depends on the models' degree of detail and to what extent they describe the actual pump type.

Fig.13.3 shows the components in a pump which cause mechanical and hydraulic losses. It involves bearings, shaft seal, front and rear cavity seal, inlet, impeller and volute casing or return channel. Throughout the rest of the chapter this Figure is used for illustrating where each type of loss occurs.

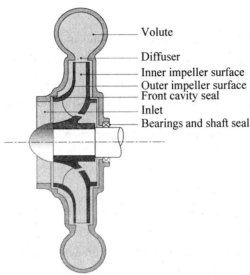

Fig.13.3 Loss causing components

13.2 Mechanical Losses

The pump coupling or drive consists of bearings, shaft seals, gear, depending on pump type. These components all cause mechanical friction losses. The following deals with losses in the bearings and shaft seals.

Bearing and shaft seal losses - also called parasitic losses - are caused by friction. They are often modeled as a constant which is added to the power consumption. The size of the losses can, however, vary with pressure and rotational speed.

The following model estimates the increased power demand due to losses in bearings and shaft seal:

$$P_{\text{loss, mechanical}} = P_{\text{loss, bearing}} + P_{\text{loss, shaft seal}} = \text{constant} \tag{13.1}$$

where

$P_{\text{loss, mechanical}}$—Increased power demand because of mechanical loss (W);

$P_{\text{loss, bearing}}$—Power lost in bearings (W);

$P_{\text{loss, shaft seal}}$—Power lost in shaft seal (W).

13.3 Hydraulic Losses

Hydraulic losses arise on the fluid path through the pump. The losses occur because of friction or because the fluid must change direction and velocity on its path through the pump. This is due to cross-section changes and the passage through the rotating impeller. The following sections describe the individual hydraulic losses depending on how they arise.

13.3.1 Flow Friction

Flow friction occurs where the fluid is in contact with the rotating impeller surfaces and the interior surfaces in the pump casing. The flow friction causes a pressure loss which reduces the head. The magnitude of the friction loss depends on the roughness of the surface and the fluid velocity relative to the surface.

13.3.2 Mixing Loss at Cross-section Expansion

For an ideal flow, the sum of pressure energy, velocity energy and potential energy is constant (Bernoulli's equation), so velocity energy is transformed to static pressure energy at cross-section expansions in the pump. The conversion is associated with a mixing loss.

The reason is that velocity differences occur when the cross-section expands, see Fig.13.4. The Figure shows a diffuser with a sudden expansion because all water particles no longer move at the same speed, friction occurs between the molecules in the fluid which results in a discharge head loss.

Even though the velocity profile after the cross-section expansion gradually is evened out, see Fig.13.4, a part of the velocity energy is turned into heat energy instead of static pressure energy.

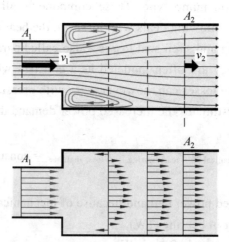

Fig.13.4　Mixing loss at cross-section expansion shown for a sudden expansion

Mixing loss occurs at different places in the pump: at the outlet of the impeller where the fluid flows into the volute casing or return channel as well as in the diffuser.

When designing the hydraulic components, it is important to create small and smooth cross-section expansions as possible.

13.3.3 Mixing Loss at Cross-section Contraction

Head loss at cross-section contraction occurs as a consequence of eddies being created in the flow when it comes close to the geometry edges, see Fig.13.5.

It is said that the flow separates. The reason for this is that the flow because of the local pressure gradients no longer adheres in parallel to the surface but instead will follow curved streamlines. This means that the effective cross-section area which the flow experiences is reduced. It is said that a contraction is made. The contraction with the area A_0 is marked on Fig.13.5. The contraction accelerates the flow and it must therefore subsequently decelerate again to fill the cross-section. A mixing loss occurs in this process. Head loss as a consequence of cross-section contraction occurs typically at the inlet to a pipe and at the impeller eye. The magnitude of the loss can be considerably reduced by rounding the inlet edges and thereby suppress separation. If the inlet is adequately rounded off, the loss is insignificant. Loss related to cross-section contraction is typically of minor importance.

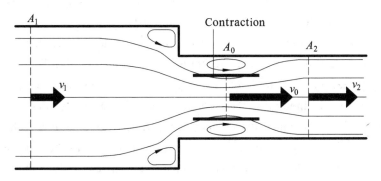

Fig.13.5 Loss at cross-section contraction

13.3.4 Recirculation Loss

Recirculation zones in the hydraulic components typically occur at part load when the flow is below the design flow. Fig.13.6 shows an example of recirculation in the impeller. The recirculation zones reduce the effective cross-section area which the flow experiences. High velocity gradients occur in the flow between the main flow which has high velocity and the eddies which have a velocity close to zero. The result is a considerable mixing loss.

Recirculation zones can occur in inlet, impeller, return channel or volute casing. The extent of the zones depends on the geometry and the operating point. When designing hydraulic components, it is important to minimize the size of the recirculation zones in the primary operating points.

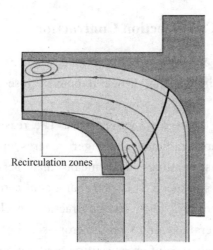

Fig.13.6   Example of recirculation in impeller

13.3.5   Incidence Loss

Incidence loss occurs when there is a difference between the flow angle and the blade angle at the impeller or guide vane leading edges. This is typically the case at part load or when pre-rotation exists.

A recirculation zone occurs on one side of the blade when there is difference between the flow angle and the blade angle, see Fig.13.7. The recirculation zone causes a flow contraction after the blade leading edge. The flow must once again decelerate after the contraction to fill the entire blade channel and mixing loss occurs.

At off-design flow, incidence losses also occur at the volute tongue. The designer must therefore make sure that flow angles and blade angles match each other so the incidence loss is minimized. Rounding blade edges and volute casing tongue can reduce the incidence loss.

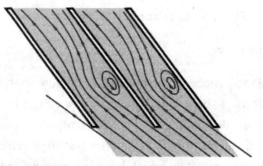

Fig.13.7   Incidence loss at inlet to impeller or guide vanes

13.3.6   Disk Friction

Disk friction is the increased power consumption which occurs on the shroud and hub of the impeller because it rotates in a fluid-filled pump casing. The fluid in the cavity between impeller and pump casing starts to rotate and creates a primary vortex. The rotation velocity equals the impeller's at the surface of the impeller, while it is zero at the surface of the pump casing. The average velocity of the primary vortex is therefore assumed to be equal to one half of the rotational velocity.

The centrifugal force creates a secondary vortex movement because of the difference in rotation velocity between the fluid at the surfaces of the impeller and the fluid at the pump casing, see Fig.13.8. The secondary vortex increases the disk friction because it transfers energy from the impeller surface to the surface of the pump casing.

The size of the disk friction depends primarily on the speed, the impeller diameter as well as the dimensions of the pump housing in particular the distance between impeller and pump casing. The impeller and pump housing surface roughness has, furthermore, a decisive importance for the size of the disk friction. The disk friction is also increased if there are rises or dents on the outer surface of the impeller e.g. balancing blocks or balancing holes.

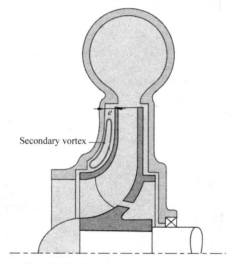

Fig.13.8 Disk friction on impeller

13.3.7 Leakage

Leakage loss occurs because of smaller circulation through gaps between the rotating and fixed parts of the pump. Leakage loss results in a loss in efficiency because the flow in the impeller is increased compared to the flow through the entire pump:

$$Q_{\text{impeller}} = Q + Q_{\text{leakage}} \tag{13.2}$$

where

Q_{impeller}—Flow through impeller (m^3/s);

Q—Flow through pump (m^3/s);

Q_{leakage}—Leakage flow (m^3/s).

Leakage occurs at many different places in the pump and depends on the pump type. Fig.13.9 shows where leakage typically occurs. The pressure differences in the pump drives the leakage flow as shown in Fig.13.10.

The leakage between the impeller and the casing at impeller eye and through axial relief are typically of the same size. The leakage flow between guide vane and shaft in multi-stage pumps are less important because both pressure difference and gap area are smaller.

To minimize the leakage flow, it is important to make the gaps as small as possible. When the pressure difference across the gap is large, it is very important that the gaps are small.

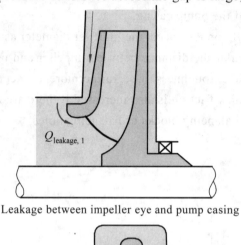
Leakage between impeller eye and pump casing

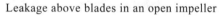
Leakage above blades in an open impeller

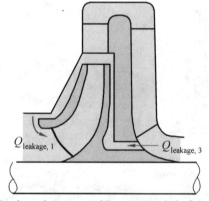
Leakage between guide vanes and shaft in a multi-stage pump

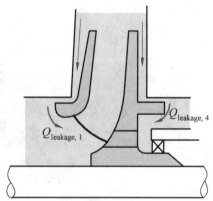
Leakage as a result of balancing holes

Fig.13.9　Types of leakage

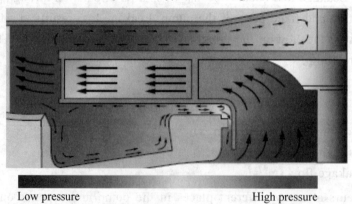

Low pressure　　　　　　　　　　　　High pressure

Fig.13.10　Leakage is driven by the pressure difference across the impeller

13.4 Loss Distribution as Function of Specific Speed

The ratio between the described mechanical and hydraulic losses depends on the specific speed n_s, which describes the shape of the impeller. Fig.13.11 shows how the losses are distributed at the design point (Ludwig et al., 2002).

Flow friction and mixing loss are significant for all specific speeds and are the dominant loss type for higher specific speeds (semi-axial and axial impellers). For pumps with low n_s (radial impellers) leakage and disk friction on the hub and shroud of the impeller will in general result in considerable losses.

At off-design operation, incidence and recirculation losses will occur.

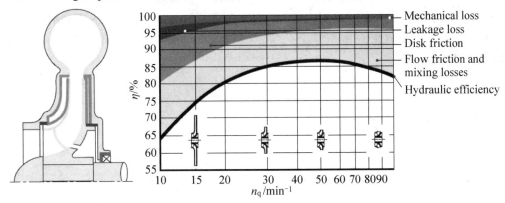

Fig.13.11 Loss distribution in a centrifugal pump as function of specific speed n_s

Questions

1. Classify hydraulic losses and mechanical losses in pumps.
2. Point out the locations of various kinds of losses on an impeller.
3. Describe the effects of various losses.
4. Explain the conditions for the occurrence of different losses.

Reference

[1] GRUNDFOS Management A/S. The Centrifugal Pump[M]. Bjerringbro: GRUNDFOS Research and Technology, 2009.

Important Words and Phrases

adhere 依附；坚持；黏着
axial relief 轴向隙角
balancing block 平衡块
balancing hole 平衡孔
bearing loss 轴承损失
Bernoulli's equation 伯努利方程
blade channel 叶道

cavity seal 腔密封
cavity 空腔；洞
coupling 联轴器；耦合
dent 凹痕，凹部
disk friction loss 圆盘摩擦损失
dominant 占优势的；支配的
eddy 涡流；旋涡

empirical 经验的
flow friction loss 流动摩擦损失
gear 齿轮
geometry edge 几何边缘；几何边界
head loss 扬程损失；水头损失
heat energy 热能
hydraulic loss 水力损失
incidence loss 冲击损失
leading edge 前缘
leakage loss 渗漏损失；容积损失
mixing loss 混合损失
molecule 分子；微小颗粒
off-design 非设计工况

parasitic loss 附加损失
pressure gradient 压力梯度
primary vortex 主涡
recirculation zone 回流区
rise 高地；增加；上升
rounding 使变圆；凑整
secondary vortex 二次涡
shaft seal loss 轴封损失
surface roughness 表面粗糙度
velocity gradient 速度梯度
velocity profile 速度（流速）剖面
volute tongue 蜗壳舌部；蜗壳隔舌

14 Characteristics of Centrifugal Pumps and Their Systems

Centrifugal pumps are one of the most common components found in fluid systems. In order to understand how a fluid system containing a centrifugal pump operates, it is necessary to understand the head and flow relationships of a centrifugal pump.

14.1 Pump Characteristic Curve

14.1.1 Theoretical Characteristic Curve

With the assumption of no whirl component of velocity at the entry to the impeller of a pump, the work done on the fluid per unit weight by the impeller is given by Eq.14.1.

$$\text{Work done on the fluid per unit weight} = v_{w2}U_2/g \tag{14.1}$$

Considering the fluid to be frictionless, the head developed by the pump will be considered the same as the theoretical head developed. Therefore the theoretical head developed can be written as

$$H_{theo} = \frac{v_{w2}U_2}{g} \tag{14.2}$$

From the outlet velocity triangle Fig.14.1

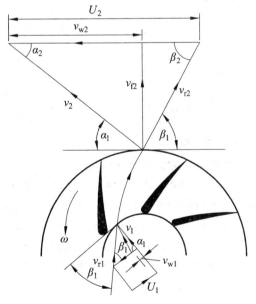

Fig.14.1 Velocity triangles for centrifugal pump Impeller

$$v_{w2} = U_2 - v_{f2}\cot\beta_2 = U_2 - (Q/A)\cot\beta_2 \tag{14.3}$$

where Q is the rate of flow at impeller outlet and A is the flow area at the periphery of the impeller.

The blade speed at outlet U_2 can be expressed in terms of rotational speed of the impeller n as $U_2 = \pi D n$.

Using this relation and the relation given by Eq.14.3, the expression of theoretical head developed can be written from Eq.14.2 as

$$H_{theo} = \pi^2 D^2 n^2 - \left[\frac{\pi D n}{A}\cot\beta_2\right]Q = K_1 - K_2 Q \tag{14.4}$$

where $K_1 = \pi^2 D^2 n^2$ and $K_2 = (\pi D n / A)\cot\beta_2$

For a given impeller running at a constant rotational speed, K_1 and K_2 are constants, and therefore head and discharge bears a linear relationship as shown by Eq.14.4. This linear variation of H_{theo} with Q is plotted as curve I in Fig.14.2.

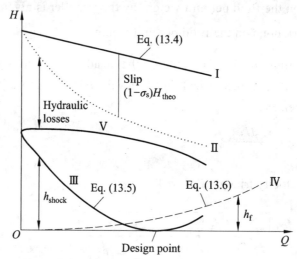

Fig.14.2 Head-discharge characteristics of a centrifugal pump

14.1.2 Actual Characteristic Curve

In course of flow through the impeller passage, there occurs a difference in pressure and velocity between the leading and trailing faces of the impeller blades. On the leading face of a blade there is relatively a high pressure and low velocity, while on the trailing face, the pressure is lower and hence the velocity is higher. This results in a circulation around the blade and a non-uniform velocity distribution at any radius. The angle at which the fluid leaves the impeller usually is smaller than the actual blade angle. This condition is called slip. The slip factor σ_S is defined as the difference between the actual and the ideal tangential velocity components at the impeller outlet.

If slip is taken into account, the theoretical head will be reduced to $\sigma_S v_{w2} U_2 / g$. Moreover the slip will increase with the increase in flow rate Q. The effect of slip in head-discharge relationship is shown by the curve II in Fig.14.2. The loss due to slip can occur in both real and ideal fluid, but in a real fluid the shock losses at entry to the blades, and the friction losses in the flow passages have to be considered.

At the design point the shock losses are zero since the fluid moves tangentially onto the blade, but on either side of the design point the head loss due to shock increases according to the relation

$$h_{shock} = K_3(Q_f - Q)^2 \tag{14.5}$$

where Q_f is the off design flow rate and K_3 is a constant.

The losses due to friction can usually be expressed as

$$h_f = K_4 Q^2 \tag{14.6}$$

where K_4 is a constant.

Eqs.14.5 and 14.6 are also shown in Fig.14.2 (curves III and IV) as the characteristics of losses in a centrifugal pump. By subtracting the sum of the losses from the head in consideration of the slip, at any flow rate (by subtracting the sum of ordinates of the curves III and IV from the ordinate of the curve II at all values of the abscissa), we get the curve V which represents the relationship of the actual head with the flow rate, and is known as the head-discharge characteristic curve of the pump.

For a centrifugal pump driven at a constant rotational speed, the head H, the power absorbed P and hence the efficiency η, as well as the $(NPSH_R)$, are functions of the flow rate Q. The relationship between these different values is represented by characteristic curves. Fig.14.3 shows an example of four characteristic curves of a single stage centrifugal pump at a rotational speed $n=1\,450$ rpm.

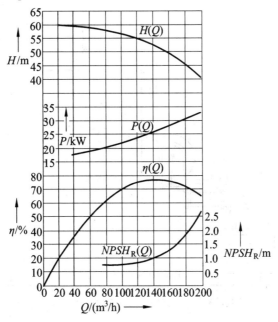

Fig.14.3 Characteristic curves of a single stage centrifugal pump

The **head / capacity curve H(Q)** also known as the **throttling curve** – represents the relationship between the head of a centrifugal pump and its flow rate. Generally the head decreases with increasing flow rate. For each value of head there exists only a single value of flow rate. The requirements for head and flow determine the overall dimensions of the pump.

The shape of the **absorbed power (power consumption) curve P(Q)** of a centrifugal pump is also a function of the flow rate, as shown in Fig.14.3. With radial flow pumps, the power absorbed increases with increasing flow rate, so they are generally started with a closed discharge valve. The power consumption is used for dimensioning of the installations which must supply the pump with energy.

The **efficiency curve** $\eta(Q)$ increases from zero with increasing flow rate to a maximum (η_{opt}) and falls as the flow rate increases further. Unless other considerations are the determining parameter for the pump selection, then the pump with optimum efficiency η_{opt} which is as close as possible to the required flow rate Q_r is selected, i.e. $Q_r \approx Q_{opt}$. The efficiency curve is used for choosing the most efficient pump in the specified operating range.

$NPSH_R$ is an abbreviation for "Net Positive Suction Head Required". $NPSH_R$ is plotted against the flow rate. **$NPSH_R$ curve** is a flat curve till the BEP of the pump and then it rises sharply beyond the best efficiency point. Net Positive Suction Head Available must be greater than the Net Positive Suction Head Required to avoid cavitation of the pump. *NPSH* Available is calculated based on the friction losses in the system while *NPSH* Required is specified by a pump vendor.

14.2 System Characteristic Curve

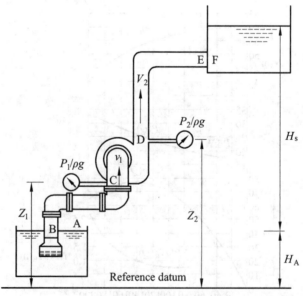

Fig.14.4 A general pumping system

Let us consider the pump and the piping system as shown in Fig.14.4. Since the flow is highly turbulent, the losses in piping system are proportional to the square of flow velocities and can, therefore, be expressed in terms of constant loss coefficients. Therefore, the losses in both the suction and delivery sides can be written as

$$h_1 = fl_1v_1^2/2gd_1 + K_1v_1^2/2g \qquad (14.7a)$$

$$h_2 = fl_2v_2^2/2gd_2 + K_2v_2^2/2g \qquad (14.7b)$$

where

h_1—the loss of head in suction side;
h_2—the loss of head in delivery side;
f—the Darcy's friction factor;
l_1, d_1 and l_2, d_2 — the lengths and diameters of the suction and delivery pipes respectively;
v_1 and v_2—accordingly the average flow velocities.

The first terms in Eqs.14.7a and 14.7b represent the ordinary friction loss (loss due to friction between fluid and the pipe wall), while the second terms represent the sum of all the minor losses through the loss coefficients K_1 and K_2 which include losses due to valves and pipe bends, entry and exit losses, etc. Therefore the total head the pump has to develop in order to supply the fluid from the lower to the upper reservoir is

$$H = H_S + h_1 + h_2 \qquad (14.8)$$

Now the flow rate through the system is proportional to the flow velocity. Therefore the resistance to flow in the form of losses is proportional to the square of the flow rate and is usually written as

$$h_1 + h_2 = \text{system resistance} = KQ^2 \qquad (14.9)$$

where K is a constant which includes, the lengths and diameters of the pipes and the various loss coefficients. The system resistance as expressed by Eq.14.9, is a measure of the loss of head at any particular flow rate through the system. If any parameter in the system is changed, such as adjusting a valve opening, or inserting a new bend, etc., then K will change. Therefore, the total head of Eqs.14.7 becomes,

$$H = H_S + KQ^2 \qquad (14.10)$$

The head H can be considered as the total opposing head of the pumping system that must be overcome for the fluid to be pumped from the lower to the upper reservoir.

The Eq.14.9 is the equation for system characteristic, and while plotted on H-Q plane (Fig.14.5), represents the system characteristic curve.

It should be noted that if there is no rise in static head of the liquid (for example pumping in a horizontal pipeline between two reservoirs at the same elevation), H_S is zero and the system head curve passes through the origin.

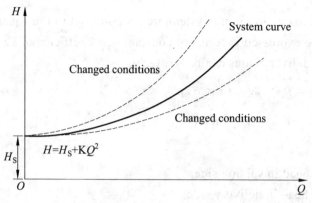

Fig.14.5 Typical system head loss curve

14.3 Matching of Pump and System Characteristics

The design point of a hydraulic pump corresponds to a situation where the overall efficiency of operation is maximum. However the exact operating point of a pump, in practice, is determined from the matching of pump characteristic with the head loss-flow, characteristic of the external system (i.e. pipe network, valve and so on) to which the pump is connected. By graphing a system characteristic curve and the pump characteristic curve on the same coordinate system, the point at which the pump must operate is identified.

The point of intersection between the system characteristic and the pump characteristic on H-Q plane is the operating point which may or may not lie at the design point that corresponds to maximum efficiency of the pump as shown in Fig.14.6. The closeness of the operating and design points depends on how good an estimate of the expected system losses has been made. The operating point shall be within the preferred operating range (POR) or the allowable operating range (AOR).

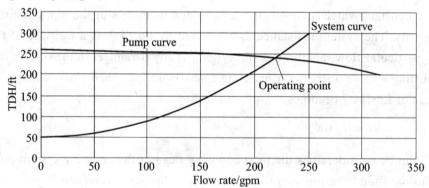

Fig.14.6 Pump and system curves

Allowable Operating Range – (AOR) is that range of flows recommended by the pump manufacturer in which the service life of the pump is not seriously compromised. Preferred Operating Range – (POR) is that range of flows specified by the customer around the pumps best efficiency capacity. Continued operation within the preferred operating range extends the pump life further. A recommend value for POR is between 70% and 120% of the BEP flow.

Questions

1. Derive the theoretical head and actual head of a centrifugal pump.
2. Draw the characteristic curves of a centrifugal pump and explain the variation of H, P, $NPSH_R$ and η with the flow rate.
3. Derive the total head of a system with centrifugal pumps.
4. Find the operating point of a centrifugal pump in a system.

References

[1] U.S. Department of Energy. The Thermodynamics, Heat Transfer, and Fluid Flow Fundamentals Handbook Volume 3 of 3: DOE-HDBK-1012/3-92[S/OL]. Virginia: National Technical Information Service, U.S. Department of Commerce, 1992: FSC-6910 [2016-08-02]. http://wenku.baidu.com/view/7fbad329bd64783e09122b1d.html?re=view.

[2] U.S. Department of Energy. DOE Fundamentals Handbook, Mechanical Science, Volume 1 of 2, Module 3 – Pumps: DOE-HDBK-1018/1-93 [S/OL]. Virginia: National Technical Information Service, U.S. Department of Commerce, 1993 [2016-06-16]. http://energy.gov/sites/prod/files/2013/06/f2/h1018v1.pdf.

[3] Sterling SIHI. Basic principles for the design of centrifugal pump installations[M]. Zoetermeer: Sterling Fluid Systems Group, 2003.

[4] NPTEL. Module 6 Pumps [EB/OL]. [2016-06-16]. http://nptel.ac.in/courses/Webcourse-contents/IIT-KANPUR/machine/ui/Course_home-8.htm.

Important Words and Phrases

abscissa 横坐标
absorbed power 吸收功率
allowable operating range 允许工作范围
BEP 高效点
best efficiency point 高效点
characteristic curve 特性曲线
Darcy's friction factor 达西摩擦因子
frictionless 无摩擦的；光滑的
head-discharge characteristic curve 扬程-流量特性曲线
intersection 交叉点
Net Positive Suction Head Available 允许气蚀余量
Net Positive Suction Head Required 必要气蚀余量

$NPSH_R$ 必要气蚀余量
optimum efficiency 最优效率
ordinate 纵坐标
outlet velocity triangle 出口速度三角形
pipe wall 管壁
preferred operating range 允许工作范围
pump characteristic curve 泵特性曲线；水泵性能曲线
service life 使用寿命
square 平方
system characteristic curve 系统特性曲线
theoretical head 理论扬程
throttling curve 节流曲线
whirl component 旋转分量；旋转组件

15 Centrifugal Pumps Operating in Systems

This chapter explains how pumps operate in a system and how they can be regulated. A pump is always connected to a system where it must circulate or lift fluid. The energy added to the fluid by the pump is partly lost as friction in the pipe system or used to increase the head.

Implementing a pump into a system results in a common operating point. If several pumps are combined in the same application, the pump curve for the system can be found by adding up the pumps' curves either serial or parallel. Regulated pumps adjust to the system by changing the rotational speed. The regulation of speed is especially used in heating systems where the need for heat depends on the ambient temperature, and in water supply systems where the demand for water varies with the consumer opening and closing the tap.

15.1 Pumps Operating in Parallel

In systems with large variations in flow and a request for constant pressure, two or more pumps can be connected in parallel. When operating the pumps, it is possible to regulate between one or more pumps at the same time. To avoid bypass circulation in pumps which are not running, a non-return valve is connected in series with each of the pumps. Fig.15.1 depicts two identical centrifugal pumps operating at the same speed in parallel.

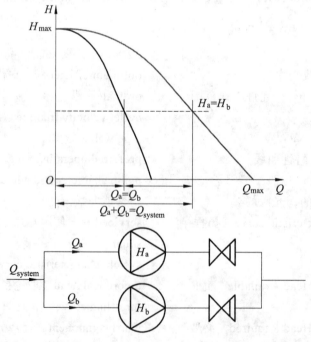

Fig.15.1 Pump characteristic curve for two identical centrifugal pumps used in parallel

Since the inlet and the outlet of each pump shown in Fig.15.1 are at identical points in the system, each pump must produce the same pump head. The total flow rate in the system, however, is the sum of the individual flow rates for each pump. The characteristics of a parallel-connected system are found by adding the single characteristics for each pump horizontally, see Fig.15.1.

When the system characteristic curve is considered with the curve for pumps in parallel, the operating point at the intersection of the two curves represents a higher volumetric flow rate than for a single pump and a greater system head loss. As shown in Fig.15.2, a greater system head loss occurs with the increased fluid velocity resulting from the increased volumetric flow rate. Because of the greater system head, the volumetric flow rate is actually less than twice the flow rate achieved by using a single pump.

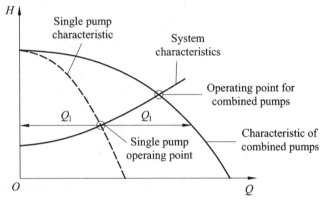

Fig.15.2 Operating point for two parallel centrifugal pumps

15.2 Pumps Operating in Series

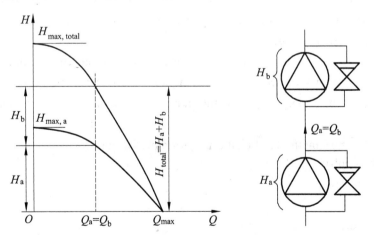

Fig.15.3 Pump characteristic curve for two identical centrifugal pumps used in series

Centrifugal pumps are used in series to overcome a larger system head loss than one pump can compensate for individually. If one of the pumps in a serial connection is not operating, it will cause a considerable resistance to the system. To avoid this, a bypass with a

133

non-return valve could be build-in. As illustrated in Fig.15.3, two identical centrifugal pumps operating at the same speed with the same volumetric flow rate contribute the same pump head. Since the inlet to the second pump is the outlet of the first pump, the head produced by both pumps is the sum of the individual heads. The volumetric flow rate from the inlet of the first pump to the outlet of the second remains the same.

As shown in Fig.15.4, using two pumps in series does not actually double the resistance to flow in the system. The two pumps provide adequate pump head for the new system and also maintain a slightly higher volumetric flow rate.

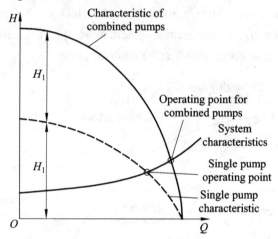

Fig.15.4 Operating point for two centrifugal pumps in series

15.3 Regulation of Pumps

When selecting a pump for a given application it is important to choose one where the duty point is in the high-efficiency area of the pump. Otherwise, the power consumption of the pump will be unnecessarily high.

However, sometimes it is not possible to select a pump that fits the optimum duty point because the requirements of the system change or the system curve changes over time. Therefore, it can be necessary to adjust the pump performance so that it meets the changed requirements.

The most common methods of changing pump performance are:
① Throttle control;
② Bypass control;
③ Speed control;
④ Impeller diameter modification.

Choosing a method of adjusting the pump performance is based on an evaluation of the initial investment together with the operating costs of the pump. All methods can be carried out continuously during operation apart from the impeller diameter modification method. Usually, oversized pumps are selected for the system and therefore it is necessary to limit the performance – first of all, the flow rate and in some applications the maximum head.

15.3.1 Throttle Control

Installing a throttle valve in serial with the pump can change the system characteristic, see Fig.15.5. The resistance in the entire system can be regulated by changing the valve settings and thereby adjusting the flow as needed. The steepness of the system characteristic curve increases so that the intersection with the characteristic curve of the pump occurs at a lower flow rate.

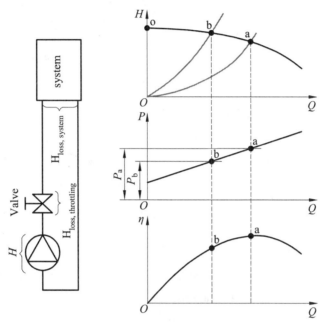

Fig.15.5 Changing of system characteristics through throttle regulation

The throttling valve causes energy losses so that continuous operation with a control valve is inefficient. The throttling losses are at a minimum when the pump characteristic curve is flat. For this reason, throttling control is mainly used for radial pumps, as with these pumps the absorbed power reduces along with the flow rate. Even where throttling control appears to be an attractive method from the aspect of initial investment costs of the control system, the total economics should be examined, especially in cases of high installed power.

For mixed flow pumps and for axial flow pumps, it should be remembered that the power absorbed increases with the decreasing flow rate. In addition axial flow pumps may move into a range of unstable operation due to the throttling process. This means rough running and an increased noise level, both being features of pumps with high specific speed, so this range of operation has to be avoided for continuous operation.

15.3.2 Bypass Control

A bypass valve is a regulation valve installed parallel to the pump, see Fig.15.6. The bypass valve guide part of the flow back to the suction line and consequently reduces the head. With a bypass valve, the pump delivers a specific flow even though the system is completely cut off. Like the throttle valve, it is possible to reduce the power consumption in some case.

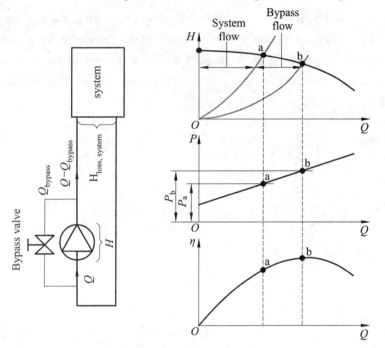

Fig.15.6 Changing of system characteristics through bypass regulation

From an overall perspective neither regulation with throttle valve nor bypass valve are an energy efficient solution and should be avoided.

15.3.3 Speed Control

The variable speed control method by means of a frequency converter is with no doubt the most efficient way of adjusting pump performance exposed to variable flow requirements. The virtue of this method is that it reduces the energy input to the system instead of dumping the excess.

When the speed of the pump is changed, the flow rate Q, the head developed H, and the power P also change. The conversion in speed is made by means of the affinity laws. These laws state that the flow rate or capacity is directly proportional to the pump speed; the discharge head is directly proportional to the square of the pump speed; and the power required by the pump motor is directly proportional to the cube of the pump speed. These laws are summarized in the following equations.

$$Q \propto n, \quad H \propto n^2, \quad P \propto n^3 \tag{15.1}$$

where
    n—speed of pump impeller (rpm);

Q—volumetric flow rate of pump (gpm or ft³/hr);
H—head developed by pump (psid or feet);
P—pump power (kW).

Using these proportionalities, it is possible to develop equations relating the condition at one speed to those at a different speed for a same pump.

$$Q_1\left(\frac{n_2}{n_1}\right) = Q_2, \quad H_1\left(\frac{n_2}{n_1}\right)^2 = H_2, \quad P_1\left(\frac{n_2}{n_1}\right)^3 = P_2 \tag{15.2}$$

which gives

$$\frac{H_2}{H_1} = \frac{Q_2^2}{Q_1^2} \quad \text{or} \quad H \propto Q^2 \tag{15.3}$$

Eq.15.3 implies that all corresponding or similar points on the Head-Discharge characteristic curves at different speeds lie on a parabola passing through the origin. All similar points on a parabola have the same efficiency and the same specific speed. So these parabolas are called constant efficiency curves or constant specific speed curves.

It is possible to develop the characteristic curve for the new speed of a pump based on the curve for its original speed. The technique is to take several points on the original curve and apply the pump laws to determine the new heads and flows at the new speed. Let A, B, C are three points on the characteristic curve (Fig.15.7) at speed n_1. For points A, B and C, the corresponding heads and flows at a new speed n_2 are found as follows:

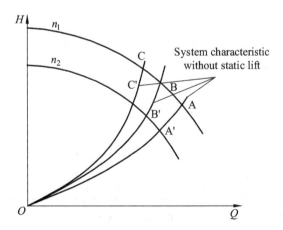

Fig.15.7 Effect of speed variation on operating point of a centrifugal pump

If the requested performance of the system changes, we can adjust the speed of the pump to allow the pump to work within the preferred operating range.

15.3.4 Impeller Diameter Modification

Reducing the impeller's diameter is a permanent change and the method can be used where the change in system demand is not temporary. The method may be energy efficient if the motor is changed and the energy consumption reduced. The change in power consumption, head and volume rate can be estimated with the help of the affinity laws, as shown in Fig.15.8.

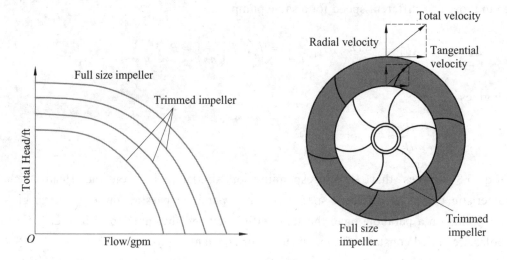

Fig.15.8 Pump curves of trimmed impeller

For a constant speed,

$$\frac{Q_1}{Q_2}=\frac{D_1}{D_2},\quad \frac{H_1}{H_2}=\left(\frac{D_1}{D_2}\right)^2,\quad \frac{P_1}{P_2}=\left(\frac{D_1}{D_2}\right)^3 \tag{15.4}$$

where

D_1, D_2 —diameters of the impeller before and after trimming (mm);

Q_1, Q_2 —volumetric flow rates of the pump before and after impeller trimming (ft³/h);

H_1, H_2 —heads developed by the pump before and after impeller trimming (feet);

P_1, P_2 —pump powers before and after impeller trimming (kW).

Trimming involves machining the impeller to reduce its diameter. Trimming should be limited to about 20% of a pump's maximum impeller diameter, because excessive trimming can result in a mismatched impeller and casing. As the impeller diameter decreases, added clearance between the impeller and the fixed pump casing increases internal flow recirculation, causes head loss, and lowers pumping efficiency, as shown in Fig.15.9.

Trimming reduces the impeller's tip speed, which in turn reduces the amount of energy imparted to the pumped fluid; as a result, the pump's flow rate and pressure both decrease. A smaller or trimmed impeller can thus be used efficiently in applications in which the current impeller is producing excessive head. In practice, impeller trimming is typically used to avoid throttling losses associated with control valves, and the system flow rate will not be affected.

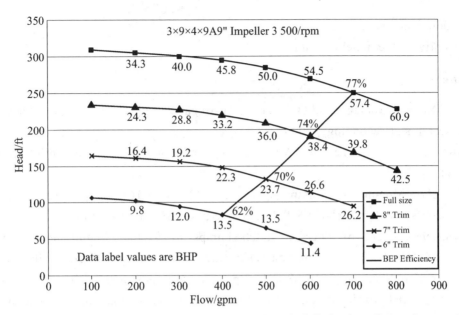

Fig.15.9 Pump curves of a centrifugal pump with a 9-inch, full size impeller and several trims

15.3.5 Comparison of Adjustment Methods

Both the throttling and the bypass method introduce some hydraulic power losses in the valves. Therefore, the resulting efficiency of the pumping system is reduced. Reducing the impeller size within the limit of 20% does not have a significant impact on the pump efficiency. Therefore, this method does not have a negative influence on the overall efficiency of the system. The efficiency of speed-controlled pumps is only affected to a limited extent, as long as the speed reduction does not drop below 50% of the nominal speed.

Each method has its pros and cons which have to be taken into account when choosing an adjustment method for a system. When it comes to obtaining the best possible efficiency, the impeller diameter adjustment method or the speed control method of the pump are the best suited for reducing the flow in the installation. When the pump has to operate in a fixed, modified duty point, the impeller diameter modification method is the best solution. However, when we deal with an installation, where the flow demand varies, the speed-controlled pump is the best solution.

Questions

1. Describe the conditions of using pumps in parallel and in series.

2. Explain the principles of throttle control and bypass control.

3. Calculate the new volumetric flow rate, head, or power for a variable speed centrifugal pump using the affinity laws.

4. Describe the effect on system flow and pump head for the following changes:
   · Changing pump speeds
   · Adding pumps in parallel

- Adding pumps in series
- Modifying the impeller diameter

References

[1] GRUNDFOS Management A/S. The Centrifugal Pump[M]. Bjerringbro: GRUNDFOS Research and Technology, 2009.

[2] GRUNDFOS Management A/S. Pump handbook[M]. Bjerringbro: GRUNDFOS Industry, 2005.

[3] NPTEL. Module 6 Pumps [EB/OL]. [2016-06-16]. http://nptel.ac.in/courses/Webcourse-contents/IIT-KANPUR/machine/ui/Course_home-8.htm.

[4] U.S. Department of Energy. Match Pumps to System Requirements: DOE/GO-102005-2160[S/OL]. Washington, DC: U.S. Department of Energy, 2005: Pumping Systems Tip Sheet #6 [2016-08-02]. http://www.nrel.gov/docs/fy06osti/38517.pdf

[5] Sterling SIHI. Basic principles for the design of centrifugal pump installations[M]. Zoetermeer: Sterling Fluid Systems Group, 2003.

[6] Centrifugal-Pump.ORG. Introduction to Affinity Laws [EB/OL]. [2016-06-16]. http://centrifugalpump.org/affinity_laws.html.

[7] Codecogs. Principles of Similarity [EB/OL]. (2011-11-3) [2016-06-16]. http://www.codecogs.com/library/engineering/fluid_mechanics/machines/pumps/principles-of-similarity.php.

[8] The Engineeirng Toolbox. Centrifugal Pumps [EB/OL]. [2016-06-16]. http://www.engineeringtoolbox.com/centrifugal-pumps-d_54.html.

Important Words and Phrases

affinity law　相似定律
ambient temperature　环境温度；周围温度
build-in　内置
bypass circulation　旁路循环
bypass control　旁路控制
bypass valve　旁通阀
circulate　使循环
compensate　补偿
cons　缺点
constant efficiency curve　等效率曲线
constant specific speed curve　等比转速曲线
cube　立方
depict　描述
dump　扔弃
duty point　工作点

energy efficient　节能的
frequency converter　变频器
horizontally　水平地
identical　完全相同的
impeller diameter　叶轮直径
impeller trimming　叶轮切削
in parallel　并联
initial investment　初期投资；最初投资
lift　提升
motor　电动机
noise level　噪声水平
nominal speed　额定转速
non-return valve　止回阀
operating cost　运营成本
origin　原点

oversized 过大的，极大的
parabola 抛物线
parallel-connected system 并联系统
proportionalities 比例
pros 优点
rough running 不平稳运转
speed control 调速控制
speed-controlled pump 调速泵

steepness 倾斜度；陡度
tap 水龙头
throttle control 节流控制
throttle valve 节流阀
tip 尖端
trimmed impeller 切削的叶轮
trimming 切削；修剪
virtue 优点

16 Cavitation and *NPSH* in Centrifugal Pumps

To understand cavitation, you must first understand vapor pressure. The vapor pressure of a liquid is the absolute pressure at which the liquid vaporizes or converts into gas at a specific temperature. Normally, the units are expressed in pounds per square inch absolute (psia). The vapor pressure of a liquid increases with its temperature, as shown in Fig.16.1. For this reason the temperature should be specified for a declared vapor pressure.

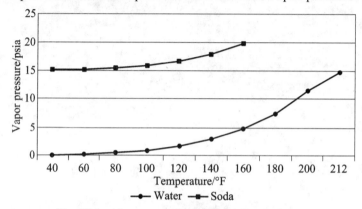

Fig.16.1 Vapor pressure versus temperature

At sea level, water normally boils at 212°F. If the pressure should increase above 14.7 psia, as in a boiler or pressure vessel, then the boiling point of the water also increases. If the pressure decreases, then the water's boiling point also decreases. For example in the Andes Mountains at 15,000 ft (4,600 meters) above sea level, normal atmospheric pressure is about 8.3 psia instead of 14.7 psia; water would boil at 184°F.

Inside the pump, the pressure decreases in the eye of the impeller because the fluid velocity increases. For this reason the liquid can boil at a lower pressure. For example, if the absolute pressure at the impeller eye should fall to 1.0 psia, then water could boil or vaporize at about 100°F.

16.1 Cavitation

16.1.1 Definition of Cavitation

The flow area at the eye of the pump impeller is usually smaller than either the flow area of the pump suction piping or the flow area through the impeller vanes. When the liquid being pumped enters the eye of a centrifugal pump, the decrease in flow area results in an increase in flow velocity accompanied by a decrease in pressure. The greater the pump flow rate, the greater the pressure drop between the pump suction and the eye of the impeller.

If the pressure drop is large enough, or if the temperature is high enough, the pressure drop may be sufficient to cause the liquid to flash to vapor when the local pressure falls below the saturation pressure for the fluid being pumped. All vapor bubbles formed by the pressure drop at the eye of the impeller are swept along the impeller vanes by the flow of the fluid. When the bubbles enter a region where local pressure is greater than saturation pressure farther out on the impeller vane, the vapor bubbles abruptly collapse.

When the vapor bubbles collapse with enough frequency, it sounds like marbles and rocks are moving through the pump. If the vapor bubbles collapse with enough energy, they can remove metal from the internal casing wall, and leave indent marks appearing like blows from a large ball pein hammer. This process of the formation and subsequent collapse or implosion of vapor bubbles in a pump is called cavitation, as shown in Fig.16.2.

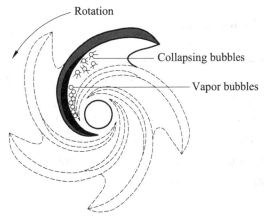

Fig.16.2　The process of cavitation

16.1.2　Consequences of Cavitation

Cavitation in a centrifugal pump has a significant effect on pump performance. Cavitation degrades the performance of a pump and reduces its efficiency, resulting in a fluctuating flow rate and discharge pressure. Flow is reduced as the liquid is displaced by vapor, and mechanical imbalance occurs as the impeller passages are partially filled with lighter vapors. This will lead to vibration and shaft deflection, eventually resulting in bearing failures, packing or seal leakage, and shaft breakage. In multi-stage pumps this can cause loss of thrust balance and thrust bearing failures.

Cavitation can also be destructive to pump's internal components. When a pump cavitates, vapor bubbles form in the low pressure region directly behind the rotating impeller vanes. These vapor bubbles then move toward the oncoming impeller vane, where they collapse and cause a physical shock to the leading edge of the impeller vane. This physical shock creates small pits on the leading edge of the impeller vane, and the imploding bubbles remove segments of impeller material. Each individual pit is microscopic in size, but the cumulative effect of millions of these pits formed over a period of hours or days can literally destroy a pump impeller, as shown in Fig.16.3.

Fig.16.3 An impeller damaged by cavitation

A small number of centrifugal pumps are designed to operate under conditions where cavitation is unavoidable. These pumps must be specially designed and maintained to withstand the small amount of cavitation that occurs during their operation. Most centrifugal pumps are not designed to withstand sustained cavitation.

16.1.3 Indications of Cavitation

Noise and vibration are indications that a centrifugal pump is cavitating. If the pump operates under cavitating conditions for enough time, the following can occur:
- Fluctuating discharge pressure, flow rate, and pump motor current.
- Pitting marks on the impeller blades and on the internal volute casing wall of the pump.
- Premature bearing failure.
- Shaft breakage and other fatigue failures in the pump.
- Premature mechanical seal failure.

16.2 Net Positive Suction Head

To avoid cavitation in centrifugal pumps, the pressure of the fluid at all points within the pump must remain above saturation pressure. The quantity used to determine if the pressure of the liquid being pumped is adequate to avoid cavitation is the net positive suction head (*NPSH*). *NPSH* is what the pump needs, the minimum requirement to perform its duties. Therefore, *NPSH* is what happens in the suction side of the pump, including what goes on in the eye of the impeller. *NPSH* takes into consideration the suction piping and connections, the elevation and absolute pressure of the fluid in the suction piping, the velocity of the fluid and the temperature. In simple terms we could say that *NPSH* is the reason that the suction nozzle is generally larger than the discharge nozzle.

To express the quantity of energy available in the liquid entering into the pump, the unit of measure for *NPSH* is feet of head or elevation in the pump suction. The pump has its $NPSH_R$, or Net Positive Suction Head Required. The system, meaning all pipes, tanks and connections on the suction side of the pump has the $NPSH_A$, or the Net Positive Suction Head Available. There should always be more $NPSH_A$ in the system than the $NPSH_R$ of the pump.

16.2.1 $NPSH_A$

The net positive suction head available ($NPSH_A$) is the difference between the pressure at the suction of the pump and the saturation pressure for the liquid being pumped. $NPSH_A$ is a characteristic of the system and must be calculated. It is determined by the plant designer and is based upon the conditions of the liquid handled, the pump location and altitude, the friction in the suction line, etc.

A formula for $NPSH_A$ can be stated as the following equation.

$$NPSH_A = h_A \pm h_Z - h_F + h_v - h_{vp}$$

Definitions and notes for different parameters in the calculation of $NPSH_A$ are listed in Table 16.1. The schematic view of different suction conditions are shown in Fig.16.4

Table 16.1 Definition and notes of different parameters in calculation of $NPSH_A$

Term	Definition	Notes
h_A	The absolute pressure on the surface of the liquid in the supply tank	· Typically atmospheric pressure (vented supply tank), but can be different for closed tanks · Don't forget that altitude affects atmospheric pressure. · Always positive (may be low, but even vacuum vessels are at a positive absolute pressure)
h_Z	The vertical distance between the surface of the liquid in the supply tank and the centerline of the pump	· Can be positive when liquid level is above the centerline of the pump (called static head) · Can be negative when liquid level is below the centerline of the pump (called suction lift) · Always be sure to use the lowest liquid level allowed in the tank
h_F	Friction losses in the suction piping	· Piping and fittings act as a restriction, working against liquid as it flows towards the pump inlet
h_v	Velocity head at the pump suction port	· Often not included as it's normally quite small
h_{vp}	Absolute vapor pressure of the liquid at the pumping temperature	· Must be subtracted in the end to make sure that the inlet pressure stays above the vapor pressure · Remember, as temperature goes up, so does the vapor pressure

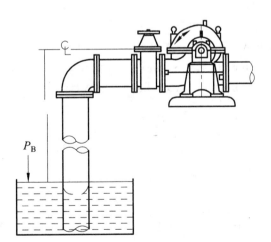

(a) Suction supply open to atmosphere with suction lift

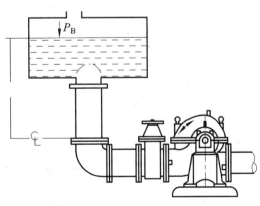

(b) Suction supply open to atmosphere with suction head

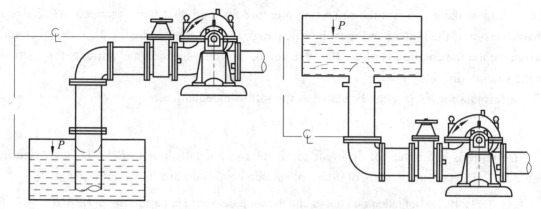

(c) Closed suction supply with suction lift   (d) Closed suction supply with suction head

Fig.16.4 Schematic view of different suction conditions

16.2.2 $NPSH_R$

The net positive suction head required ($NPSH_R$) is the energy in the liquid required to overcome the friction losses from the suction nozzle to the eye of the impeller without causing vaporization (cavitation). It is a characteristic of the pump and is indicated on the pump's curve. It depends upon factors including type of impeller inlet, impeller design, pump flow rate, impeller rotational speed, and the type of liquid being pumped. It is determined by a lift test, producing a negative pressure in inches of mercury and converted into feet of required $NPSH$.

According to the Standards of the Hydraulic Institute, a suction lift test is performed on the pump and the pressure in the suction vessel is lowered to the point where the pump suffers a 3% loss in total head. This point is called the $NPSH_R$ of the pump. Some pump manufacturers perform a similar test by closing a suction valve on a test pump and other manufacturers lower the suction elevation.

If you want to know the $NPSH_R$ of your pump, the easiest method is to read it on your pump curve. It's a number that changes normally with a change in flow. When the $NPSH_R$ is mentioned in pump literature, it is normally the value at the best efficiency point. Then, you'll be interested in knowing exactly where your pump is operating on its curve. A typical $NPSH_R$ characteristic curve is shown in Fig.16.5.

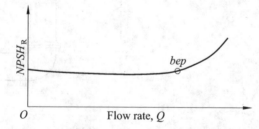

Fig.16.5 A typical $NPSH_R$ characteristic curve

If you don't have your pump curve, you can determine the $NPSH_R$ of your pump with the

following formula:
$$NPSH_R = ATM + Pgs + h_v - h_{vp}$$
where

> ATM—the atmospheric pressure at the elevation of the installation expressed in feet of head.
>
> Pgs—the suction pressure gauge reading taken at the pump centerline and converted into feet of head.
>
> h_v—Velocity head = $v^2/2g$, where, v is the velocity of the fluid moving through the pipes measured in feet per second, and g is the acceleration of gravity (32.16 ft/sec);
>
> h_{vp}—the vapor pressure of the fluid expressed in feet of head. The vapor pressure is tied to the fluid temperature.

The condition that must exist to avoid cavitation is that the net positive suction head available must be greater than or equal to the net positive suction head required. That is to say there must be more suction side pressure available than the pump requires. This requirement can be stated mathematically as shown below.

$$NPSH_A \geqslant NPSH_R$$

No engineer wants to be responsible for installing a noisy, slow, damaged pump. It's critical to get the $NPSH_R$ value from the pump manufacturer and to insure that your $NPSH_A$ pressure will be adequate to cover that requirement.

16.3 Preventing Cavitation

If a centrifugal pump is cavitating, several changes in the system design or operation may be necessary to increase the $NPSH_A$ above the $NPSH_R$ and stop the cavitation. One method for increasing the $NPSH_A$ is to increase the pressure at the suction of the pump. For example, if a pump is taking suction from an enclosed tank, either raising the level of the liquid in the tank or increasing the pressure in the space above the liquid increases the suction pressure.

It is also possible to increase the $NPSH_A$ by decreasing the temperature of the liquid being pumped. Decreasing the temperature of the liquid decreases the saturation pressure, causing $NPSH_A$ to increase.

If the head losses in the pump suction piping can be reduced, the $NPSH_A$ will be increased. Various methods for reducing head losses include increasing the pipe diameter, reducing the number of elbows, valves, and fittings in the pipe, and decreasing the length of the pipe.

It may also be possible to stop cavitation by reducing the $NPSH_R$ for the pump. The $NPSH_R$ is not a constant for a given pump under all conditions, but depends on certain factors. Typically, the $NPSH_R$ of a pump increases significantly as flow rate through the pump increases.

Therefore, reducing the flow rate through a pump by throttling a discharge valve decreases $NPSH_R$. $NPSH_R$ is also dependent upon pump speed. The faster the impeller of a pump rotates, the greater the $NPSH_R$ is. Therefore, if the speed of a variable speed centrifugal pump is reduced, the $NPSH_R$ of the pump decreases. However, since a pump's flow rate is most often dictated by the needs of the system on which it is connected, only limited adjustments can be made without starting additional parallel pumps, if available.

Questions

1. Define the terms Net Positive Suction Head and cavitation.
2. List the consequences of cavitation.
3. Calculate the $NPSH_R$ and $NPSH_A$.
4. List factors affecting the $NPSH_R$ and $NPSH_A$.
5. Provide measures in preventing cavitation.

References

[1] U.S. Department of Energy. DOE Fundamentals Handbook, Mechanical Science, Volume 1 of 2, Module 3 – Pumps: DOE-HDBK-1018/1-93 [S/OL]. Virginia: National Technical Information Service, U.S. Department of Commerce, 1993 [2016-06-16]. http://energy.gov/sites/prod/files/2013/06/f2/h1018v1.pdf.

[2] Bachus, Larry, and Angel Custodio. Know and understand centrifugal pumps[M]. Amsterdam: Elsevier, 2003.

[3] Centrif. One Hour Centrifugal Pump University [EB/OL]. [2016-06-16]. http://www.pricepump.com/pumpschool/Column_With_Contents.htm.

[4] Messina J P. Pump handbook [M]. New York: McGraw-Hill, 1986.

[5] Magdy Abou Rayan, Nabil H. Mostafa, Purage Ohans. Textbook of Machines Hydraulic [M]. Zagazig: Zagazig University, Dept. of Mechanical Engineering, 2009.

[6] Sahdev M. Centrifugal Pumps: Basic Concepts of Operation, Maintenance, and Troubleshooting[J]. Part I, Presented at The Chemical Engineers' Resource Page, www. cheresources. com, 2004: 1-28.

[7] Christopher Earls Brennen. Hydrodynamics of Pumps [EB/OL]. (2000-01-01) [2016-06-16]. http://authors.library.caltech.edu/25019/1/content.htm.

Important Words and Phrases

absolute pressure 绝对压力，绝对压强	boiling point 沸点
Andes Mountains 安第斯山脉	breakage 破坏；破损
ball pein hammer 圆头铁锤	cavitate 形成空洞；成穴，空化
blow 打击	collapse 溃灭
boil 煮沸，沸腾	cumulative effect 累积效应
boiler 锅炉；烧水壶，热水器	declared 宣布的

deflection   挠曲变形
degrade   降低
destructive   破坏的；具有破坏性的
dictate   控制，支配
displace   取代；置换
enclosed   封闭的
flash   闪蒸
flow velocity   流速
fluctuate   波动；涨落
friction loss   摩擦损失
implode   向内破裂；内爆
implosion   向内破裂；内爆
indent   凹痕
indication   迹象；象征
lift test   提升测试
literally   不夸张地
marble   大理石

microscopic   微小的；微观的
pitting mark   点蚀痕
pounds per square inch absolute   磅/平方英寸（绝对压强）
premature   过早的；提前的
pressure gauge   压力计，测压表
pressure vessel   压力容器
psia   磅/平方英寸（绝对压强）
saturation pressure   饱和压力
sea level   海平面
soda   苏打水；碳酸水
suction lift   吸水高度；吸升水头
suction piping   吸水管系统
suction port   吸入口
vaporization   蒸发
vaporize   使……蒸发；蒸发
vented   通风的；开孔的

17 Axial flow Pumps

Axial flow pumps are one of three subtypes of centrifugal pumps, the others being mixed flow and radial flow. Of these three types, axial flow pumps are characterized by the highest flow rates and lowest discharge pressures. An axial flow pump may generate only 10 to 20 feet of head, much lower than most other types of centrifugal pumps. They are capable of producing very high flow rates – as high as several hundred thousand gallons per minute, the highest flow rates of any type of centrifugal pump.

Axial-flow pumps differ from radial-flow in that the fluid enters and exits along the same direction parallel to the rotating shaft. The fluid is not accelerated but instead "lifted" by the action of the impeller. The impeller looks and operates similar to a boat propeller, which is the reason why axial flow pumps are also called propeller pumps.

17.1 Basic Structure of an Axial Flow Pump

An axial flow pump is a common type of pump that essentially consists of a propeller (an axial impeller) in a pipe. The diameter of the pump casing is nearly the same as that of the suction chamber. It can be installed vertically, horizontally or obliquely. An axial flow pump is basically composed by the following parts: suction chamber, impeller (including blades and hub), guide vanes, shaft, elbow discharge, upper and lower bearings, stuffing box, governing mechanism of blade angles, etc., as shown in Fig.17.1.

1. Suction chamber

To improve the hydraulic conditions at the inlet, the suction chamber usually has a flared inlet.

2. Impeller

The impeller consists of a central boss with a number of blades mounted on it. The usual number of impeller blades lies between 2 and 8, with a hub diameter to impeller diameter ratio of 0.3 to 0.6. The impeller rotates within a cylindrical casing with fine clearance between the blade tips and the casing walls. The impeller is driven by a motor that is either sealed directly in the pump body or by a drive shaft that enters the pump tube from the side.

The blades are oriented in such a way that the pumped fluid exits axially (i.e., in the same direction as the shaft), rather than radially (90 degrees from the shaft). Fluid particles, in course of their flow through the pump, do not change their radial locations since the change in radius at the entry (called "suction") and the exit (called "discharge") of the pump is very small. The axial orientation of the impeller vanes produces very low head as the liquid is pumped.

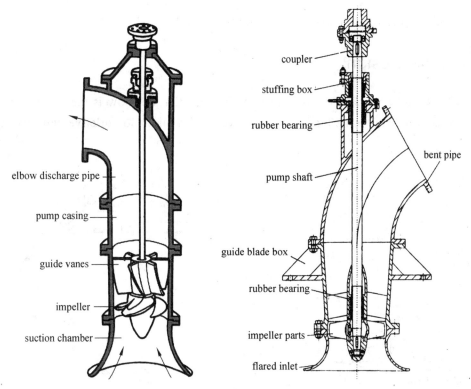

Fig.17.1 Axial flow pump structure

The impeller blades can be fixed, adjustable when stationary, or fully adjustable, as shown in Fig.17.2. For the axial flow pump with fixed blades, the blades are integrally cast with the hub, and the setting angle (pitch) of the blade can not be altered. For the axial flow pump with blades adjustable when stationary, the blades are bolted on the hub, and the setting angle of the blade can be altered, but the turbine must be stopped to do that. For the axial flow pump with fully adjustable blades, the impeller pitch can be adjusted through a set of oil pressure regulating mechanism at any time to run efficiently at different conditions.

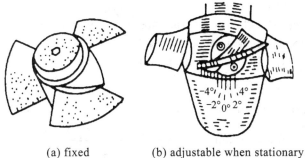

(a) fixed   (b) adjustable when stationary
Fig.17.2 Axial flow pump impeller

3. Guide vanes

The outlet guide vanes are provided to eliminate the whirling component of velocity at discharge and to convert the energy to pressure. The machine may be fitted with inlet guide vanes to eliminate pre-rotation and to make the flow purely axial.

17.2 Blade Design

Fig.17.3 shows an axial flow pump. It can be seen that the flow is the same at inlet and outlet. The section through the blade at x-x is shown with inlet and outlet velocity triangles in Fig17.4.

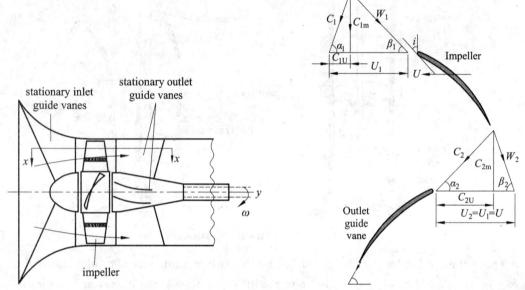

Fig.17.3 A propeller of an axial flow pump   Fig.17.4 Velocity triangles of an axial flow pump

From the centrifugal pump's Euler's equation, the impeller's head is

$$H = \frac{(U_2 \cdot C_{2U} - U_1 \cdot C_{1U})}{g} \tag{17.1}$$

where
 U —the tangential velocity;
 C_U —the projection velocity of the absolute velocity C on the direction of the tangential velocity.

If we consider the conditions at a mean radius r_m, then

$$U_2 = U_1 = U = r_m \cdot \omega \tag{17.2}$$

where ω is the angular velocity of the impeller.

Work done on the fluid per unit weight $= \dfrac{U \cdot (C_{2U} - C_{1U})}{g}$ \hfill (17.3)

For maximum energy transfer, $C_{1U} = 0$, that is, $\alpha_1 = 90°$. Again, from the outlet velocity triangle,

$$C_{2U} = U - C_{2m} \cot \beta_2 \tag{17.4}$$

where C_m is the projection velocity of the absolute velocity C on the radial direction.

Assuming a constant flow from inlet to outlet,

$$C_{1m} = C_{2m} = C_m \tag{17.5}$$

Therefore, the maximum energy transfer per unit weight by an AFP is,

$$H = U\frac{(U - C_m \cot\beta_2)}{g} \tag{17.6}$$

Fluid particles at different radii will obtain different energies from the pump. There will be energy conversion between them after flowing out of the impeller, and there will be a lot of energy losses in that process. For constant energy transfer over the entire span of the blade, the above equation should be constant for all values of r. But, U^2 will increase with an increase in radius r, therefore to maintain a constant value an equal increase in $UC_m \cot\beta_2$ must take place. Since, C_m is constant, therefore $\cot\beta_2$ must increase with increasing r. Therefore, the blade must be twisted as the radius changes, as shown in Fig.17.5.

Fig.17.5 Twisted blades of an axial flow pump

17.3 Airfoil Theory

The axial flow pump is operating on the basis of lift theory of airfoil in aerodynamics. The cross section of the impeller blade is similar to the shape of an airfoil.

According to the knowledge of fluid dynamics, the fluid will be separated into two streams when flowing around the airfoil, one above the airfoil and the other below the airfoil, as shown in Fig.17.6. The flow path above the airfoil is longer than that below the airfoil. Therefore, the upper surface of the airfoil will have large velocity and low pressure, while the lower surface will have small velocity and high pressure. The resultant force on the airfoil by the fluid is upward, that's why the plane can fly. Meanwhile, the airfoil will produce a downward reactive force on the fluid.

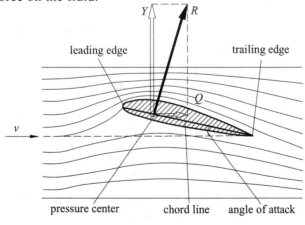

Y—lift, R—total aerodynamic force, Q—drag

Fig.17.6 Schematic view of airfoil lift

In an axial flow pump, the airfoil blades are obliquely installed on the impeller hub, as shown in Fig.17.7. The airfoil blades rotate with the hub in the fluid. The streamline of airfoil blade is contrary to that of an airplane, which means that the upper surface of the airfoil blade has small velocity and high pressure, while the lower surface has large velocity and low pressure. The resultant force from the fluid to the airfoil blades is downward, while the resultant force from the airfoil blades to the fluid is upward. Under the action of the high speed rotating airfoil blades, the liquid is lifted.

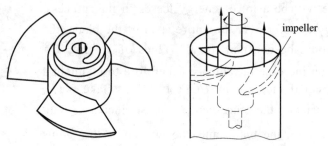

Fig.17.7 Axial flow pump impeller design and flow

The various terms (as shown in Fig.17.8) related to airfoils are defined below:

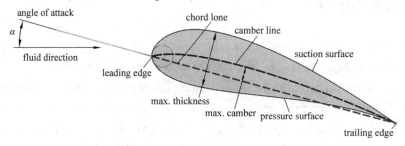

Fig.17.8 Airfoil nomenclature

· The suction surface (a.k.a upper surface) is generally associated with higher velocity and lower static pressure.

· The pressure surface (a.k.a lower surface) has a comparatively higher static pressure than the suction surface. The pressure gradient between these two surfaces contributes to the lift force generated for a given airfoil.

The geometry of the airfoil is described with a variety of terms:

· The leading edge is the point at the front of the airfoil that has maximum curvature (minimum radius).

· The trailing edge is defined similarly as the point of maximum curvature at the rear of the airfoil.

· The chord line is the straight line connecting the leading and trailing edges. The chord length, or simply chord, c, is the length of the chord line. That is the reference dimension of the airfoil section.

The shape of the airfoil is defined using the following geometrical parameters:

· The mean camber line or mean line is the locus of points midway between the upper and lower surfaces. Its shape depends on the thickness distribution along the chord.

· The thickness of an airfoil varies along the chord. It may be measured in either of two ways:

✓ Thickness measured perpendicular to the camber line. This is sometimes described as the "American convention".

✓ Thickness measured perpendicular to the chord line. This is sometimes described as the "British convention".

Finally, important concepts used to describe the airfoil's behavior when moving through a fluid are:

· The aerodynamic center, which is the chord-wise length about which the pitching moment is independent of the lift coefficient and the angle of attack.

· The center of pressure, which is the chord-wise location about which the pitching moment is zero.

17.4 Performance of an Axial Flow Pump

The performance characteristics of axial flow pumps are different from other pump types. Pump performance curves, which are provided by a manufacturer to describe the correlation between head and capacity of an individual pump, can be used to describe these characteristics. Fig.17.9 shows a typical performance curve for an axial flow pump manufactured by Batescrew.

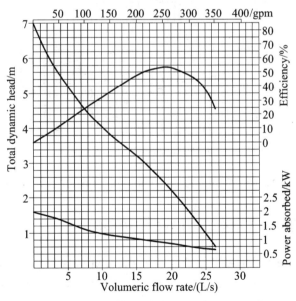

Fig.17.9 Typical performance curve for an axial flow pump

The image depicts the relationship between head, flow rate, power, and efficiency of an axial flow pump. Even though an axial flow pump produces very low heads at its normal operating point, the curve of head versus capacity is much steeper than with other centrifugal pump types. As shown in the diagram, the shut-off (zero flow) head of an axial flow pump can be as much as three times the head at the pump's best efficiency point. Additionally, the power requirement increases as flow decreases, with the highest power draw at shut off. That's why the axial flow pump can cause overloading of the drive motor if the flow rate is reduced significantly from the design capacity. These trends are opposite to radial flow centrifugal pumps, which require more power as flow rate increases.

This collection of curves in Fig.17.10 shows the change in performance at different blade pitch (angle). The upper lines show the different performance at varying blade pitch, the lower lines indicate the absorbed power. The power requirements and pump head increases with an increase in pitch, thus allowing the pump to adjust according to the system conditions to provide the most efficient operation.

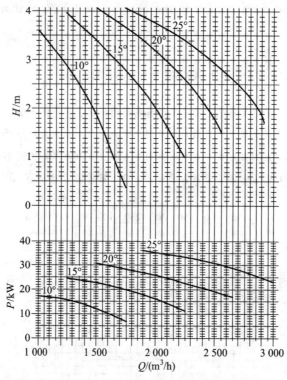

Fig.17.10 Characteristic curve of an axial-flow pump

17.5 Advantages of Axial Flow Pumps

The main advantage of an AFP is that it has a relatively high discharge (flow rate) at a relative low head. For example, it can pump up to 3 times more water and other fluids at lifts of less than 4 meters as compared to the more common radial-flow or centrifugal pump. It also can be easily be adjusted to run at peak efficiency at low-flow/high-pressure and high-flow/low-pressure by changing the pitch on the propeller.

The effect of turning the fluid is not too severe in an axial pump and the length of the impeller blades is also short. This leads to lower aerodynamic losses and higher stage efficiencies. These pumps have the smallest of the dimensions among many of the conventional pumps and are more suited for low heads and higher discharges.

17.6 Applications of Axial Flow Pumps

Axial flow pumps are used in applications that require very high flow rates and very low amounts of pressure (head). Axial flow pumps are ideal for a wide range of applications such as:

- filling or emptying of reservoirs;
- flood and storm water control;
- circulation of large quantities of water in power plants or chemical industries;
- large volume drainage and irrigation;
- raw water intake;
- transfer of liquids in large-scale municipal sewage treatment plants;
- water level control in coastal and low-lying areas;
- filling and emptying of dry docks and harbor installations.

Questions

1. State the purposes of the following components in an axial flow pump:
   - Suction chamber
   - Impeller
   - Guide vanes
2. Explain the reason for twisted blades of an axial flow pump.
3. Illustrate the airfoil theory.
4. Compare the performance curves of a radial flow pump and an axial flow pump.

References

[1] NPTEL. Module 6 Pumps [EB/OL]. [2016-06-16]. http://nptel.ac.in/courses/Webcourse-contents/IIT-KANPUR/machine/ui/Course_home-8.htm.

[2] Engineering 360. Axial Flow Pumps Information [EB/OL]. [2016-06-16]. http://www.globalspec.com/learnmore/flow_transfer_control/pumps/axial_flow_pumps.

[3] Codecogs. Axial Flow Pumps and Fans [EB/OL]. (2011-11-23) [2016-06-16]. http://www.codecogs.com/library/engineering/aerodynamics/axial-flow-pumps-and-fans.php.

Important Words and Phrases

aerodynamic center 空气动力中心   aerodynamic loss 空气动力损失
aerodynamic force   空气动力   aerodynamics 空气动力学

airfoil blade　翼型叶片
airfoil theory　翼形理论
airfoil　翼型；机翼
angle of attack　迎角
boss　轮毂；套筒
camber line　弧线
cast　浇铸
chord line　弦线
coupler　联轴器
curvature　曲率
drainage　排水
drive motor　驱动电动机
drive shaft　驱动轴；主动轴
dry dock　干船坞
elbow discharge　出水弯管
flared　喇叭形；向外展开的
fluid dynamics　流体动力学
fluid particle　流体质点；流体粒子
geometry　几何结构
harbor installation　港口设施；海港设备
inlet guide vane　进口导叶
intake　取水（口）
irrigation　灌溉
lift theory　提升理论
locus　轨迹
low-lying area　低洼地区
mean line　等分线

municipal　市政的
obliquely　倾斜地
operating point　工况点
oriented　确定……的方位
outlet guide vane　出口导叶
overload　超载；过载
pitch　节距；斜度，度
pitching moment　俯仰力矩
pressure surface　压力面；叶片正面
projection velocity　投影速度
propeller　螺旋桨
radii　半径（radius 的复数）
raw water　原水
resultant force　合力
setting angle　安装角；装置角
severe　严重的；剧烈的
sewage treatment plant　污水处理厂
shut-off head　关闭扬程
span　跨度
stage efficiency　分级效率
subtype　子类型
suction chamber　吸入室
suction surface　吸力面；叶片背面
trailing edge　后缘；机翼后缘
twisted　扭曲的
whirling component　旋转分量

18 Positive Displacement Pumps

18.1 Introduction

A positive displacement pump is one in which a definite volume of liquid is delivered for each cycle of pump operation. This volume is constant regardless of the resistance to flow offered by the system that the pump is in, provided the capacity of the power unit driving the pump or pump component strength limits are not exceeded. The positive displacement pump delivers liquid in separate volumes with no delivery in between, although a pump having several chambers may have an overlapping delivery among individual chambers, which minimizes this effect. The positive displacement pump differs from centrifugal pumps, which deliver a continuous flow for any given pump speed and discharge resistance.

Positive displacement pumps can be grouped into three basic categories based on their design and operation. The three groups are reciprocating pumps, rotary pumps, and diaphragm pumps.

18.2 Principle of Operation

All positive displacement pumps operate on the same basic principle. This principle can be most easily demonstrated by considering a reciprocating positive displacement pump consisting of a single reciprocating piston in a cylinder with a single suction port and a single discharge port as shown in Fig.18.1. Check valves in the suction and discharge ports allow flow in only one direction.

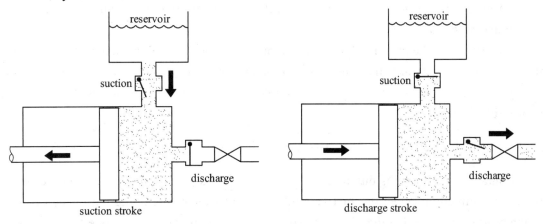

Fig.18.1 Reciprocating positive displacement pump operation

During the suction stroke, the piston moves to the left, causing the check valve in the suction line between the reservoir and the pump cylinder to open and admit water from the reservoir.

During the discharge stroke, the piston moves to the right, sealing the check valve in the suction line and opening the check valve in the discharge line. The volume of liquid moved by the pump in one cycle (one suction stroke and one discharge stroke) is equal to the change in the liquid volume of the cylinder as the piston moves from its farthest left position to its farthest right position.

18.3 Reciprocating Pumps

Reciprocating positive displacement pumps are generally categorized in four ways: direct-acting or indirect-acting; simplex or duplex; single-acting or double-acting; and power pumps.

18.3.1 Direct-Acting and Indirect-Acting Pumps

Some reciprocating pumps are powered by prime movers that also have reciprocating motion, such as a reciprocating pump powered by a reciprocating steam piston. The piston rod of the steam piston may be directly connected to the liquid piston of the pump or it may be indirectly connected with a beam or linkage. Direct-acting pumps have a plunger on the liquid (pump) end that is directly driven by the pump rod (also the piston rod or extension thereof) and carries the piston of the power end. Indirect-acting pumps are driven by means of a beam or linkage connected to and actuated by the power piston rod of a separate reciprocating engine.

18.3.2 Simplex and Duplex Pumps

A simplex pump, sometimes referred to as a single pump, is a pump having a single liquid (pump) cylinder. A duplex pump is the equivalent of two simplex pumps placed side by side on the same foundation.

The driving of the pistons of a duplex pump is arranged in such a manner that when one piston is on its upstroke the other piston is on its downstroke, and vice versa. This arrangement doubles the capacity of the duplex pump compared to a simplex pump of comparable design.

18.3.3 Single-Acting and Double-Acting Pumps

A single-acting pump is one that takes a suction, filling the pump cylinder on the stroke in only one direction, called the suction stroke, and then forces the liquid out of the cylinder on the return stroke, called the discharge stroke. A double-acting pump is one that, as it fills one end of the liquid cylinder, is discharging liquid from the other end of the cylinder. On the return stroke, the end of the cylinder just emptied is filled, and the end just filled is emptied. One possible arrangement for single-acting and double-acting pumps is shown in Fig.18.2.

18.3.4 Power Pumps

Power pumps convert rotary motion to low speed reciprocating motion by reduction gearing, a crankshaft, connecting rods and crossheads. Plungers or pistons are driven by the crosshead drives. Rod and piston construction, similar to duplex double-acting steam pumps, is used by the liquid ends of the low pressure, higher capacity units. The higher pressure units are normally single-acting plungers, and usually employ three (triplex) plungers. Three or more plungers substantially reduce flow pulsations relative to simplex and even duplex pumps.

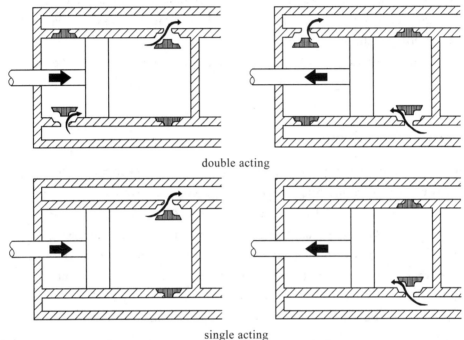

Fig.18.2 Single-acting and double-acting pumps

Power pumps typically have high efficiency and are capable of developing very high pressures. They can be driven by either electric motors or turbines. They are relatively expensive pumps and can rarely be justified on the basis of efficiency over centrifugal pumps. However, they are frequently justified over steam reciprocating pumps where continuous duty service is needed due to the high steam requirements of direct-acting steam pumps.

In general, the effective flow rate of reciprocating pumps decreases as the viscosity of the fluid being pumped increases because the speed of the pump must be reduced. In contrast to centrifugal pumps, the differential pressure generated by reciprocating pumps is independent of fluid density. It is dependent entirely on the amount of force exerted on the piston.

18.4 Rotary Pumps

Rotary pumps operate on the principle that a rotating vane, screw, or gear traps the liquid in the suction side of the pump casing and forces it to the discharge side of the casing. These pumps are essentially self-priming due to their capability of removing air from suction lines and producing a high suction lift. In pumps designed for systems requiring high suction lift and self-priming features, it is essential that all clearances between rotating parts, and between rotating and stationary parts, are kept to a minimum in order to reduce slippage. Slippage is the leakage of fluid from the discharge of the pump back to its suction.

Due to the close clearances in rotary pumps, it is necessary to operate these pumps at relatively low speed in order to secure reliable operation and maintain pump capacity over an extended period of time. Otherwise, the erosive action due to the high velocities of the liquid passing through the narrow clearance spaces would soon cause excessive wear and increased clearances, resulting in slippage.

There are many types of positive displacement rotary pumps, and they are normally grouped into three basic categories that include gear pumps, screw pumps, and moving vane pumps.

18.4.1 Simple Gear Pump

There are several variations of gear pumps. The simple gear pump shown in Fig.18.3 consists of two spur gears meshing together and revolving in opposite directions within a casing. Only a few thousandths of an inch clearance exists between the case and the gear faces and teeth extremities. Any liquid that fills the space bounded by two successive gear teeth and the case must follow along with the teeth as they revolve. When the gear teeth mesh with the teeth of the other gear, the space between the teeth is reduced, and the entrapped liquid is forced out the pump discharge pipe. As the gears revolve and the teeth disengage, the space again opens on the suction side of the pump, trapping new quantities of liquid and carrying it around the pump case to the discharge. As liquid is carried away from the suction side, a lower pressure is created, which draws liquid in through the suction line.

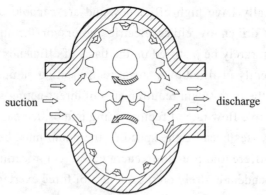

Fig.18.3 Simple gear pump

With the large number of teeth usually employed on the gears, the discharge is relatively smooth and continuous, with small quantities of liquid being delivered to the discharge line in rapid succession. If designed with fewer teeth, the space between the teeth is greater and the capacity increases for a given speed; however, the tendency toward a pulsating discharge increases. In all simple gear pumps, power is applied to the shaft of one of the gears, which transmits power to the driven gear through their meshing teeth.

There are no valves in the gear pump to cause friction losses as in the reciprocating pump. The high impeller velocities, with resultant friction losses, are not required as in the centrifugal pump. Therefore, the gear pump is well suited for handling viscous fluids such as fuel and lubricating oils.

18.4.2 Screw-Type Positive Displacement Rotary Pump

There are many variations in the design of the screw type positive displacement, rotary pump. The primary differences consist of the number of intermeshing screws involved, the pitch of the screws, and the general direction of fluid flow. Two common designs are the two-screw, low-pitch, double-flow pump and the three-screw, high-pitch, double-flow pump.

1. Two-Screw, Low-Pitch, Screw Pump

The two-screw, low-pitch, screw pump consists of two screws that mesh with close clearances, mounted on two parallel shafts. One screw has a right-handed thread, and the other screw has a left-handed thread. One shaft is the driving shaft and drives the other shaft through a set of herringbone timing gears. The gears serve to maintain clearances between the screws as they turn to promote quiet operation. The screws rotate in closely fitting duplex cylinders that have overlapping bores. All clearances are small, but there is no actual contact between the two screws or between the screws and the cylinder walls.

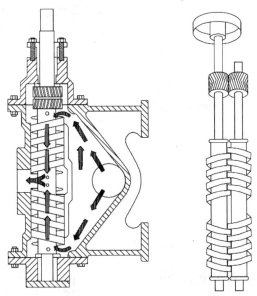

Fig.18.4　Two-screw, low-pitch, screw pump

The complete assembly and the usual flow path are shown in Fig.18.4. Liquid is trapped at the outer end of each pair of screws. As the first space between the screw threads rotates away from the opposite screw, a one-turn, spiral-shaped quantity of liquid is enclosed when the end of the screw again meshes with the opposite screw. As the screw continues to rotate, the entrapped spiral turns of liquid slide along the cylinder toward the center discharge space while the next slug is being entrapped. Each screw functions similarly, and each pair of screws discharges an equal quantity of liquid in opposed streams toward the center, thus eliminating hydraulic thrust. The removal of liquid from the suction end by the screws produces a reduction in pressure, which draws liquid through the suction line.

2. Three-Screw, High-Pitch, Screw Pump

The three-screw, high-pitch, screw pump, shown in Fig.18.5, has many of the same elements as the two-screw, low-pitch, screw pump, and their operations are similar. Three screws, oppositely threaded on each end, are employed. They rotate in a triple cylinder, the two outer bores of which overlap the center bore. The pitch of the screws is much higher than in the low pitch screw pump; therefore, the center screw, or power rotor, is used to drive the two outer idler rotors directly without external timing gears. Pedestal bearings at the base support the weight of the rotors and maintain their axial position. The liquid being pumped enters the suction opening, flows through passages around the rotor housing, and through the screws from each end, in opposed streams, toward the center discharge. This eliminates unbalanced hydraulic thrust. The screw pump is used for pumping viscous fluids, usually lubricating, hydraulic, or fuel oil.

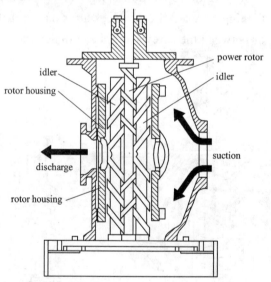

Fig.18.5 Three-screw, high-pitch, screw pump

18.4.3 Rotary Moving Vane Pump

The rotary moving vane pump shown in Fig.18.6 is another type of positive displacement pump used. The pump consists of a cylindrically bored housing with a suction inlet on one side and a discharge outlet on the other. A cylindrically shaped rotor with a diameter smaller than the cylinder is driven about an axis placed above the centerline of the cylinder. The clearance between rotor and cylinder is small at the top but increases at the bottom. The rotor carries vanes that move in and out as it rotates to maintain sealed spaces between the rotor and the cylinder wall. The vanes trap liquid or gas on the suction side and carry it to the discharge side, where contraction of the space expels it through the discharge line. The vanes may swing on pivots, or they may slide in slots in the rotor.

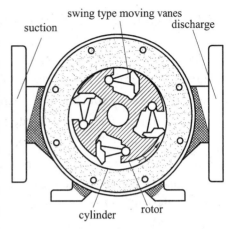

Fig.18.6　Rotary moving vane pump

18.5　Diaphragm Pumps

Diaphragm pumps are also classified as positive displacement pumps because the diaphragm acts as a limited displacement piston. The pump will function when a diaphragm is forced into reciprocating motion by mechanical linkage, compressed air, or fluid from a pulsating, external source. The pump construction eliminates any contact between the liquid being pumped and the source of energy. This eliminates the possibility of leakage, which is important when handling toxic or very expensive liquids. Disadvantages include limited head and capacity range, and the necessity of check valves in the suction and discharge nozzles. An example of a diaphragm pump is shown in Fig.18.7.

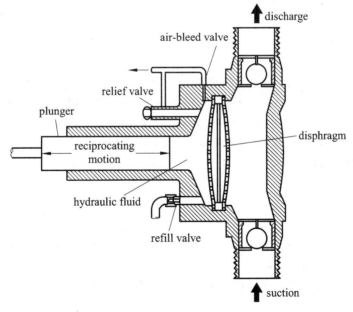

Fig.18.7　Diaphragm pump

18.6 Positive Displacement Pump Characteristic Curves

Positive displacement pumps deliver a definite volume of liquid for each cycle of pump operation. Therefore, the only factor that affects flow rate in an ideal positive displacement pump is the speed at which it operates. The flow resistance of the system in which the pump is operating will not affect the flow rate through the pump.

Fig.18.8 shows the characteristic curve for a positive displacement pump. The dashed line in Fig.18.8 shows the actual positive displacement pump performance. This line reflects the fact that as the discharge pressure of the pump increases, some amount of liquid will leak from the discharge of the pump back to the pump suction, reducing the effective flow rate of the pump. The rate at which liquid leaks from the pump discharge to its suction is called slippage.

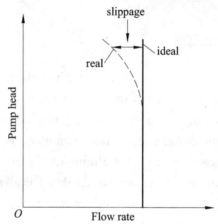

Fig.18.8　Positive displacement pump characteristic curve

18.7　Positive Displacement Pump Protection

Positive displacement pumps are normally fitted with relief valves on the upstream side of their discharge valves to protect the pump and its discharge piping from over-pressurization. Positive displacement pumps will discharge at the pressure required by the system they are supplying. The relief valve prevents system and pump damage if the pump discharge valve is shut during pump operation or if any other occurrence such as a clogged strainer blocks system flow.

Questions

1. State the difference between the flow characteristics of centrifugal and positive displacement pumps.

2. Given a simplified drawing of a positive displacement pump, classify the pump as one of the following:

　· Reciprocating piston pump

- Gear-type rotary pump
- Screw-type rotary pump
- Moving vane pump
- Diaphragm pump

3. Explain the importance of viscosity as it relates to the operation of a reciprocating positive displacement pump.

4. Describe the characteristic curve for a positive displacement pump.

5. Define the term slippage.

6. State how positive displacement pumps are protected against over-pressurization.

Reference

[1] U. S. Department of Energy. DOE Fundamentals Handbook, Mechanical Science, Volume 1 of 2, Module 3 – Pumps: DOE-HDBK-1018/1-93 [S/OL]. Virginia: National Technical Information Service, U.S. Department of Commerce, 1993 [2016-06-16]. http://energy.gov/sites/prod/files/2013/06/f2/h1018v1.pdf.

Important Words and Phrases

actuate 使动作
air-bleed valve 排气阀
chamber 腔，室
check valve 止回阀
clogged 阻塞的
crankshaft 曲轴
crosshead 十字头；联杆器
deliver 输送
diaphragm pump 隔膜泵
direct-acting 直接联动式
discharge line 排水管线
discharge port 排出口
discharge stroke 排出冲程
disengage 分开
double-acting 双作用式
downstroke 下行程；下行冲程
duplex 双缸式
entrap 夹带
exert 施加（影响、压力等）
extremity 末端；端点
fuel 燃料
gear face 齿轮端面

gear pump 齿轮泵
gear teeth 齿轮齿
herringbone 人字形的
idler 惰轮；中间齿轮
indirect-acting 间接联动式
intermeshing 相互啮合的
left-handed thread 左旋螺纹
mesh 啮合
moving vane pump 动叶片泵
occurrence 出现
overlapping bore 重叠的孔
pedestal bearing 托架轴承；支承轴承
pitch 螺距
pivot 枢轴
plunger 柱塞
positive displacement pump 容积式泵；正排量泵
power pump 动力泵
power unit 动力设备；动力装置
pulsate 有规律地跳动
reciprocating piston 往复活塞
reciprocating pump 往复泵；循环泵

reduction gearing　减速装置，减速齿轮
refill valve　加注阀
relief valve　安全阀；减压阀
revolve　旋转
right-handed thread　右旋螺纹
rotary pump　转子泵；旋转泵，回转泵
screw pump　螺杆泵
screw　螺杆
self-priming　自灌式；自吸式
simplex　单缸式
single-acting　单作用式
slippage　回流
slot　槽
slug　少量的液体
spur gear　正齿轮

steam piston　蒸汽活塞
strainer　过滤器
stroke　冲程
suction stroke　吸入冲程
swing　摆动
thousandths　千分之一的
three-screw, high-pitch, screw pump　三螺旋高螺距螺旋泵
timing gear　调速齿轮；正时齿轮
two-screw, low-pitch, screw pump　双螺旋低螺距螺旋泵
upstroke　上行程；上行冲程
viscosity　黏度
viscous fluid　黏性流体

参考译文[①]

0 绪论

0.1 透平机械定义

透平机械是指那些通过一组或多组转动叶栅的动力作用把能量传递给连续流动的液体或从流动的液体中获得能量的机械设备。"turbo"或"turbinis"来源于拉丁语，指的是旋转或回旋运动。从本质上来说，根据对机械设备的要求，转动的叶栅、转子或叶轮通过做正功或负功改变流过液体的滞止焓。焓的变化与流体中压力的变化是密切相关的，而且是同时发生的。

0.2 透平机械分类

0.2.1 根据能量转化方向

透平机械可以分为两大类：一类是通过吸收能量增加液体的压力或水头的机械设备（如风机、压缩机和泵）；另一类是通过流体扩张，降低压力或水头而产生能量的机械设备（如风轮机、水轮机、蒸汽轮机和燃气轮机）。图0.1选择性地给出了现实中常用的几种透平机械。各种各样的服务需求，促进了多种类型的泵（压缩机）或涡轮机的开发。一般来说，对于一个给定的运行需求，仅有一种泵或涡轮机最适用，能实现最优的运行工况。本书仅对水轮机和水泵进行了探讨。

0.2.2 根据流动路径

可以根据液体流过叶轮时的流动路径，对透平机械进行分类。流体流过叶轮时，流动路径完全或基本平行于旋转轴的设备被称为轴流式透平机械，如图0.1（a）与（e）所示。流动路径完全或基本垂直于旋转轴的设备被称为径流式透平机械，如图0.1（c）所示。混流式透平机械的使用范围最为广泛。本书中的混流是指在叶轮出口处的径向速度分量与轴向速度分量都占很大比例的流动。图0.1（b）为一台混流泵，图0.1（d）为一台混流式水轮机。

0.2.3 根据压力变化

根据液体流过转轮时是否存在压力变化，还可以把透平机械分为冲击式机械或反击式机械。在冲击式机械中，所有的压力变化都在一个或多个喷嘴中完成，流体是被导向叶轮的。

[①] 本书英文部分多摘自英联邦国家的技术资料，其中许多英制单位在国内无明确规范。为最大程度还原技术资料，结合其写作环境，译文中延用了大部分英制单位。下附译文中出现的部分英制单位—公制单位换算关系。
1英尺 = 12英寸 = 0.304 8米
1平方英尺 = 144平方英寸 = 0.092 9平方米
1立方英尺 = 1 728立方英寸 = 0.028 3立方米
1加仑（英）= 4.546升
1磅 = 0.453 6千克
1马力（米制）= 0.735瓦

图 0.1（f）中的切击式水轮机就是一种冲击式水轮机。在反击式机械中，当流体流过转轮时，动能和压能都改变。图 0.1（a~e）都是反击式机械。

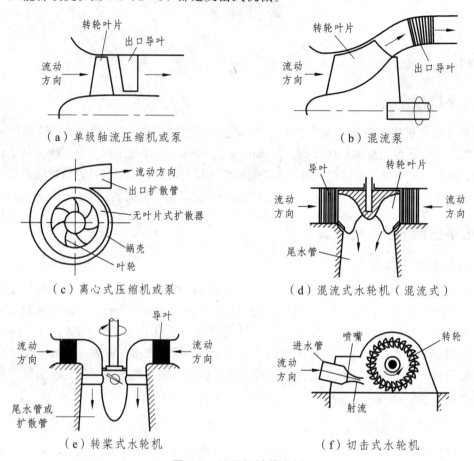

图 0.1　透平机械的实例

1 水轮机

1.1 简介

水力机械是一种把流体的机械能传递给运动部件（转轮、活塞等）或把运动部件的机械能传递给液体的机器设备。那些把流动液体的能量传递给机械的运动部件，并且液体进入时的能量比流出时的能量高的水力机械被称为水轮机。那些把运动部件的能量传递给液体，并且液体进入时的能量比流出时的能量低的水力机械被称为水泵。

从牛顿定律中可知，只有作用力才能改变流体的动量。同样，如果流体的动量改变，必定会产生作用力。该原理在水轮机中也适用。水轮机的叶片或斗叶安装在轮毂上，用来改变水流方向，从而改变水的动量。水流流过轮毂时动量改变了，产生的合力使轮毂的轴旋转并做功发电。

水轮机把水的势能和动能转换为可用的机械能。水轮机轴上可用的机械能用来驱动直接耦合在水轮机轴上的发电机。由水能产生的电能被称为水电。

1.2 发展历史

水轮机有很长的发展历史。最古老最简单的形式为水车。水车在古希腊最先开始使用，然后推广到中世纪的欧洲，主要用来研磨谷物等。大约在 1830 年，法国的工程师贝努瓦·富聂隆发明了第一台成功用于商业的水轮机。后来富聂隆又建造了适用于工业用途的水轮机，转速可以达到 2300 转/分，功率达 50 千瓦，效率高于 80%。

美国工程师詹姆斯·比切诺·法兰西斯设计了第一台径向流入式水轮机。这种水轮机被广泛使用，效果很好，而且评价很高。其原型应用的水头在 10 米至 100 米之间。图 1.1 为这种水轮机的简图，从中可以看到流动路径基本上是由径向到轴向。

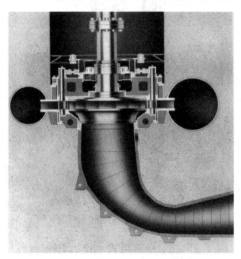

图 1.1 混流式水轮机的简图

佩尔顿（切击式）水轮机是根据其发明者莱斯特·阿伦·佩尔顿的名字来命名的，这种类型的水轮机在 19 世纪后期开始投入使用。这是一种冲击式水轮机，管道中的高压水通过喷嘴，在大气压下喷射出来。形成的射流冲击水轮机的叶片（斗叶），产生所需的扭矩和功率输出。图1.2为切击式水轮机的简图。最初的水头在90米至900之间（现代切击式水轮机的工作水头可以达到2000米）。

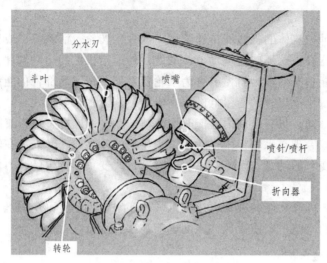

图1.2 切击式水轮机的简图

20世纪早期，为了适应不断增长的能源需求，一种新型水轮机诞生了。这种水轮机的工作水头较小，仅为3米至9米，用于可以建坝的河段上。1913年，维克多·卡普兰发明了螺旋桨式（卡普兰式）水轮机，如图1.3所示，这种水轮机工作时就像轮船上反转的螺旋桨。后来，卡普兰对这种水轮机进行了改进，使叶片可以旋转，因此提高了水轮机在主要运行条件下（即可用的流量和扬程）的效率。

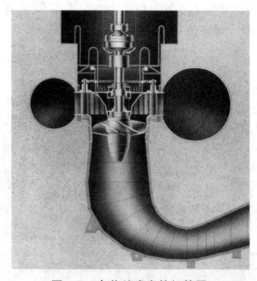

图1.3 卡普兰式水轮机简图

1.3 水轮机的分类

水轮机的叶片、桨叶或斗叶在水的作用下绕轴旋转。水轮机或水车的转动部分通常被称为转轮。水轮机的旋转运动驱动发电机发电或驱动其他旋转机械。水轮机是一种通过水的动力或压力作用形成扭矩的一种机器。

可以根据水流流过叶片时的方向、水轮机进口的水头、水轮机的比转速以及液体流过转轮时压力的变化，对水轮机进行分类。

1.3.1 根据水流流过转轮时的方向

切流式水轮机：在这种水轮机中，水流沿着与轮毂相切的方向冲击转轮。实例：切击式水轮机（见图1.2）。

径流式水轮机：在这种水轮机中，水流沿着径向流动。水流可以由外到内或由内到外径向流动。如果水流从转轮外部流向转轮内部，这种径流式水轮机被称为内流式（向心式）水轮机。实例：老式混流水轮机。如果水流从内部流向外部，这种水轮机被称为外流式水轮机。实例：富聂隆式水轮机（见图1.4）。

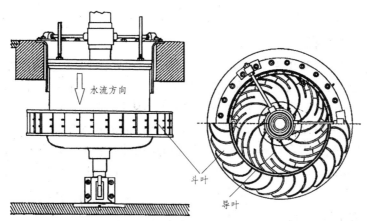

图 1.4 富聂隆式水轮机

轴流式水轮机：水流的流向与水轮机轴平行。实例：转桨式水轮机和定桨式水轮机（见图1.3）。

混流式水轮机：水流径向进入转轮，轴向流出转轮。实例：现代的混流式水轮机（见图1.1）。

1.3.2 根据水轮机进口的水头

高水头水轮机：这种水轮机的净水头在150米至2000米之间，甚至更高。这种水轮机需要的流量小。实例：切击式水轮机。

中水头水轮机：这种水轮机的净水头在30米至150米之间，需要的流量中等。实例：混流式水轮机。

低水头水轮机：这种水轮机的净水头小于30米，需要的流量很大。实例：转桨式水轮机。

1.3.3 根据水轮机主轴的布置方式

水轮机主轴可以为立式，也可以为卧式。在现代实践中，切击式水轮机通常为卧式，而其他水轮机，特别是大型机组，都采用立式布置方式。

1.3.4 根据水轮机的比转速

水轮机的比转速的定义为几何类似的水轮机在一个单位的水头（1米水头）下运转产生一个单位的功率时所需的转速。可由下面的公式表示：$n_s = \dfrac{n\sqrt{P}}{H^{5/4}}$。

低比转速水轮机：比转速小于50（单射流时在10至35之间，双射流时可以达到50）。实例：切击式水轮机。

中比转速水轮机：比转速在50至250之间。实例：混流式水轮机。

高比转速水轮机：比转速大于250。实例：转桨式水轮机。

1.3.5 根据液体流过水轮机时压力的变化

根据液体流过转轮时压力的变化，可以把水轮机分为两大类：一类是冲击式水轮机，这种类型的水轮机利用高速水射流的动能把水能转换为机械能；第二类是反击式水轮机，这种水轮机通过水的压能和动能的联合作用产生能量。

1.3.5.1 冲击式水轮机

冲击式水轮机由喷嘴喷出来的高速射流进行驱动，高速射流冲击安装在叶轮周围的水斗。产生的冲力使水轮机转动，流体的动能随即减小。在水轮机斗叶上的流体压力没有变化，所有的压降都在喷嘴处产生。在到达水轮机前，通过喷嘴的加速，流体的压头转换为速度头。由于水流在到达转轮斗叶之前通过喷嘴形成射流，所以冲击式水轮机不需要在转轮周围设置压力外壳。可以用牛顿第二定律解释冲击式水轮机中的能量转换。从本质上讲，冲击式水轮机仅转换了流体的动能，并没有改变其压力。冲击式水轮机是一个大类，还可以细分为几种，每一种的工作方式稍有区别。

1. 切击式（佩尔顿）水轮机

在切击式水轮机的转轮圆周上安装了多个杯形的斗叶，转轮与中心轮毂（圆盘）相连（见图1.2）。喷嘴布置在转轮的周边，把水流射向这些杯形的斗叶，并把水的势能转换为动能，推动水轮机的转轮旋转。

2. 斜击式水轮机

斜击式水轮机是从切击式水轮机演变而来的。斜击式水轮机的斗叶不是完整的杯状，而是半杯状（见图1.5）。与切击式水轮机相比，这些半杯状的斗叶使水流进入和流出斗叶的速度更快、流量更大，因此能量效率更高。

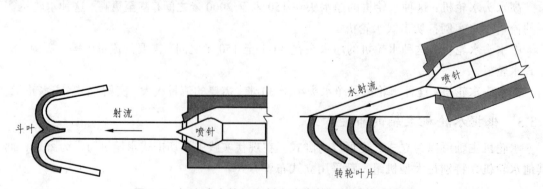

图1.5 切击式水轮机和斜击式水轮机的叶片和喷嘴

3. 双击式水轮机

双击式水轮机的叶片是放射状布置的槽形叶片,这些叶片都安装在圆筒形的转轮上(见图 1.6)。为了使水流流线更加平滑,把叶片的进水口和末端都设计成锥形的。双击式水轮机只有两个喷嘴,射流分别以 45 度的角冲击叶片,把冲击力转换为动能。控制机构可以调节流出喷嘴的流量。在这类水轮机中,水流两次流经转轮叶片,一次是从叶片外流到叶片内,另一次是从叶片内流到叶片外。通常情况下,双击式水轮机的流量比切击式水轮机大。双击式水轮机有时又称为米歇尔式-班克式或奥森博格式水轮机。

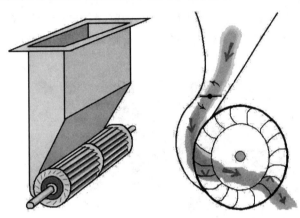

图 1.6 双击式水轮机

1.3.5.2 反击式水轮机

反击式水轮机的工作原理是反击。水流以高压低速进入水轮机的引流部件。此时一部分压能转化为动能,然后水流流入转轮,压能进一步转换为动能。随着水流流过转轮,转轮被加速,在转轮叶片上形成一个反作用力,转轮开始旋转。

由于静液压力作用在叶片的两侧,所以没有做功。所有的功都是在能量转换为动能时产生的。值得注意的是相对速度从进口到出口不断增加,而绝对速度在不断减小。

在反击式水轮机中,水轮机内的水是有压力的,工作时水轮机内是充满水的。因此,这种水轮机必须放在一个能够承压的机壳内。冲击式水轮机的机壳仅起到保护转轮、防止水外溅的功能,不具备任何水力功能。

在反击式水轮机运行时,转轮室是充满水的,尾水管用来回收尽可能多的水头。因此,这种水轮机又称为整周进水式水轮机。

水流流过反击式水轮机时的流动路径有几种。根据液体流过转轮的流动路径,可以把反击式水轮机分为三大类:混流式水轮机、轴流式水轮机和径流式水轮机。

1. 混流式水轮机

大部分水轮机都有轴向流分量和径向流分量。这种类型的水轮机被称为混流式水轮机。法兰西斯式水轮机就是一种典型的混流式水轮机。在法兰西斯式水轮机中,水流径向流入、轴向流出。

2. 轴流式水轮机

轴流式水轮机通常是指定桨式水轮机或卡普兰式(转桨式)水轮机。典型的桨式水轮机为一台立式机器,其蜗壳和径向活动导叶的构造与混流式水轮机的进水部分非常相似(见图

1.1 和 1.3）。水流径向进入水轮机，转一个直角弯，然后轴向进入转轮。

3. 径流式水轮机

在这种类型的水轮机中，水流主要在一个与旋转轴垂直的平面内流动。富聂隆式水轮机就是早期的一种径流式水轮机，水流径向流出转轮。

反击式水轮机还包括斜流式或德列阿兹式和贯流式水轮机。

4. 斜流式或德列阿兹式水轮机

斜流式水轮机（见图 1.7）填补了定桨式水轮机和混流式水轮机之间的空缺。转轮叶片以一定的角度安装在锥形轮毂的边缘。叶片外缘是开放的。叶片可以转动，叶片的轴与水轮机的轴成 45 度的倾角。叶片轴与水轮机主轴的交角随着水头的增加而降低，可以在 30 度和 60 度之间变化。

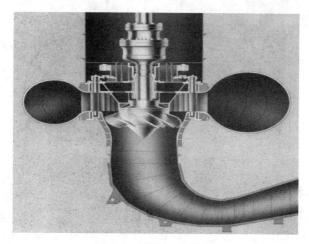

图 1.7　斜流式水轮机

5. 贯流式水轮机

贯流式水轮机（见图 1.8）通常是开发潮汐能源及水头极低流量特大的水能的最佳选择。这种水轮机的优点是流量大、流速高、开挖量少等。贯流式水轮机的引水部件、转轮和排水部件都在同一个轴线上。水流沿着直线直接流入转轮，不需要蜗壳。水流的流向从管道的进口到尾水管的出口都是轴向的。

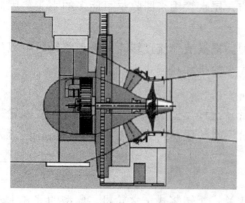 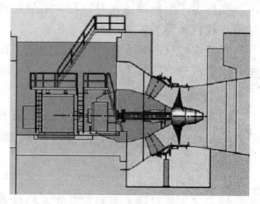

（a）灯泡式水轮机　　　　　　　　（b）竖井式水轮机

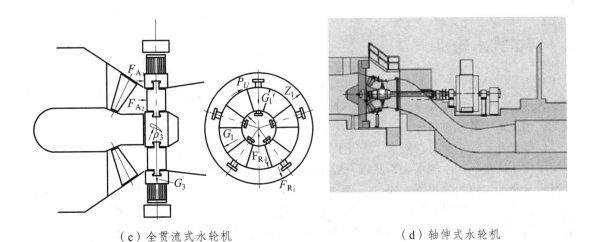

（c）全贯流式水轮机　　　　　　　　　（d）轴伸式水轮机

图 1.8　贯流式水轮机

通常把贯流式水轮机分为两大类：全贯流式和半贯流式。全贯流式水轮机的发电机转子在水轮机转轮的外圆上，由于密封比较困难，应用较少。半贯流式水轮机的发电机与水轮机是分开的。根据构造类型，半贯流式水轮机又分为灯泡式、竖井式和轴伸式。

灯泡贯流式水轮机的发电机组安装在密封的灯泡体内，结构紧凑、流道顺直、水力效率高，所以被广泛采用。竖井贯流式水轮机的发电机安装在一个竖井内。轴伸贯流式水轮机的发电机安装在流道的外面，水轮机的主轴从尾水管中伸到外面。

1.4　冲击式水轮机与反击式水轮机的区别（见表 1.1）

表 1.1 中列出了冲击式水轮机与反击式水轮机的不同之处。

表 1.1　冲击式水轮机与反击式水轮机的区别

特　征	冲击式水轮机	反击式水轮机
流体能量的转换	流体所有可用的能量都通过喷嘴转换为动能	在流体进入转轮之前，仅有一部分的流体能量转换为动能
压力和流速的变化	水流作用在转轮上的整个过程中，压力都是恒定不变的（大气压）	水流进入转轮时有多余的压力，流速和压力随着水流流过转轮都发生变化
水流对叶片的作用	叶片只有在喷嘴前面时才动作	叶片一直在动作
水流进入转轮的方式	水流可以进入一部分或整个转轮圆周	水流进入整个转轮圆周
密封的蜗壳	不需要	需要
水流充满转轮/水轮机的程度	转轮和叶片并不完全充满水或被水淹没	在水轮机运行的整个过程中，水流完全充满所有叶道
机组安装	只能安装在尾水位以上，不需要尾水管	可以安装在尾水位以上或以下，需要尾水管
水的相对速度	水流流过动叶片时，其相对速度保持不变或因摩擦而有所减小	水流流过叶道时压力在不断地下降，所以相对速度在不断地增加
流量调节	通过安装在喷嘴上的针阀调节流量	通过活动导叶调节流量

2 切击式水轮机

切击式水轮机（水斗式水轮机/佩尔顿水轮机）是一种冲击式水轮机，水流切向进入转轮，在进入点上的可用能量完全是动能。切击式水轮机更适用于水头非常高、流量和比转速都比较低的情况。水轮机进口和出口处的压力都为大气压。

水流离开转轮后所有动能都"消失了"。由于转轮在近乎大气压下运转，而且不能利用机组高于尾水高程所产生的水头，所以一般没有采用尾水管。另外对于高水头的装置，相对来说可用水头的这点损失也不重要。

切击式水轮机可以采用卧轴或立轴。通常，卧式布置适用于中小型的水轮机，仅有一个或两个射流。但有些卧式切击式水轮机设计了 4 个射流。配有多个射流的大型切击式水轮机通常采用立轴。射流对称地分布在转轮的外围以平衡水冲力。

切击式水轮机的结构复杂，包括很多组成部分。当然并不是每个造好的切击式水轮机都包括这些组成部分。这意味着一个制造商与另一个制造商制造的切击式水轮机或一个小型与一个大型切击式水轮机的结构都会不同。因此，这里仅讨论了关键的组成部分。

2.1 喷射机构（喷管）

大型的压力水管把水引入水电站，在压力水管的末端安装了一个喷射机构，把压能转换为动能，如图 2.1 所示。喷射机构把流动的液体转换成高速射流，冲击安装在转轮上的斗叶或叶片，带动水轮机转轮旋转。通过针阀前后移动控制冲击到叶片上的水量。由于水在喷嘴口和针阀之间的环形区域内流动，针阀向前移动时流量就会降低，反之亦然。

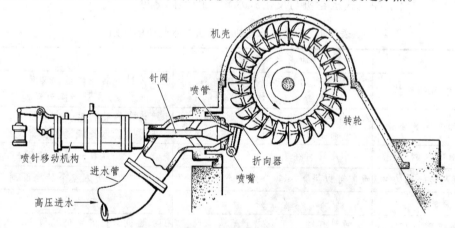

图 2.1 切击式水轮机的横截面图

喷管的作用为① 引导压力水管中的水流以合适的角度冲击转轮；② 改变水量以适应瞬时负荷，进而调节水轮机。

对于多喷嘴式水轮机，喷管位于压力水管上的弯管尾部或位于配水支管的末端。

喷管由①喷嘴、②针阀（喷针）、③折向器组成。喷针在喷嘴内同轴滑动。喷针的滑动控制喷嘴的开度，从而控制进入转轮的水量。

2.2 喷针移动机构（接力器）

喷针移动机构是用来移动喷针的。如果喷针移动机构安装在喷嘴的外面，在喷嘴直管段的前面必须设置一个弯管，使喷杆能够伸出来。在这种喷针移动机构中，水流仅在一个平面内对称，喷嘴内部的流态会造成能量损失。这种喷嘴对于中水头、中出力的机组来说效果很好，但不适用于现代的高水头、高出力的机组。而且喷管检修拆卸也很困难。因此开发了一种直流式喷射机构（直喷嘴或内控式喷射机构，见图2.2），把喷针移动机构设置在喷嘴体的内部，以保证水流顺直。这种直喷嘴效率更高、更小，结构轻巧，拆装也更容易。

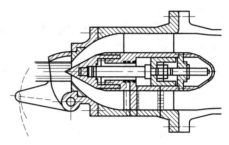

图 2.2 直喷嘴

2.3 折向器

折向器位于喷嘴出口处转轮斗叶的前面（见图2.2）。当突然出现甩负荷并且需水量降低时，喷针将迅速减小过流面积。否则水轮机将产生飞逸。但是如果关闭动作太快，长长的压力水管中流动的水柱就会突然减速，导致压力剧增，可能会使压力水管破裂。为了保证压力水管中的压力在极限范围内并且抑制可能出现的转轮飞逸，可以把折向器移动到射流的路径中，使部分或全部射流偏离转轮。随着喷嘴内喷针慢慢关闭，达到稳定状态，调节机构再把折向器移回到原来的位置。要精心设计折向器的形状以把射流切断到所需的程度。应该注意的是，折向器的径向运动不应对斗叶的转动产生干扰。

2.4 转 轮

转轮是一个安装在主轴上的圆柱形轮辐，转轮外围等间距地安装了很多斗叶，如图2.3所示。

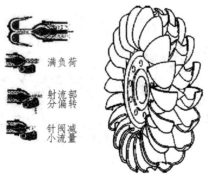

图 2.3 切击式水轮机转轮

叶片又称为斗叶，看起来像中间由分水刃连接起来的两个半椭圆的杯子。这种形状可以获得最大的效率。射流进入斗叶的中间结合处（分水刃），被分成两股，流过斗叶的工作面，然后从出水边流出。这种分流可以平衡产生的轴向推力。

射流被偏转一个168度到176度的角，这样正好避开下一个斗叶的背面。

在效率最高的情况下，射流应垂直冲击斗叶。但是这一点并不是总能实现的，因为每个斗叶拦截射流的角度都是 $2\pi/n$（式中 n 为斗叶的数量）。为了使射流尽可能更多地垂直冲击斗叶，把斗叶向半径的后面倾斜一个角度，倾角在 $10°\sim18°$。

斗叶的外缘均有一个缺口，缺口的作用是使下一个斗叶不会干扰冲击上一个斗叶的射流。

斗叶的数量由以下两个条件确定：

① 斗叶的数量必须足够多，在任何时候都可以拦截射流，在没有把能量传递给转轮之前，不允许水流直接流入尾水渠。

② 目前斗叶的数量在18至24之间，极限情况下范围是14至30之间。

2.5 喷 嘴

切击式水轮机可以采用卧式或立式的布置方式。如果轴是水平的，可以采用一个或两个喷嘴冲击转轮。有时水轮机会有两个转轮，可以并排放置或交替放置。发电机可以由安放在两侧的两个水轮机共同驱动。

在立轴的布置方式中仅有一个转轮，喷嘴的数量在2至6之间，大部分水轮机的喷嘴数为4或6。

切击式水轮机是低比转速装置。可以通过增加喷嘴数量增加比转速，比转速随着喷嘴数平方地增加而增大。但是，要求进入斗叶的射流必须相距较远，不会相互干扰。在实际应用中，射流数目的上限为6个。

2.6 配水管

为了给这些喷嘴供水，需要设置配水管。配水管是安装在压力水管末端的管段。对于卧式双喷嘴机组，配水管是Y形的，在叉管的两臂上设置弯管以安装喷管（见图2.4）。Y形的角度在70度至90度之间变化。

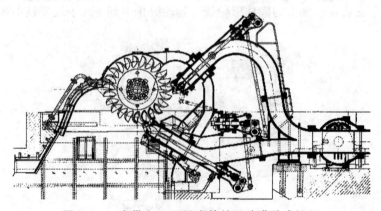

图 2.4 一台带有 Y 形配水管的双喷嘴卧式机组

对于立轴的布置方式，最简单的配水管构造为从压力水管末端开始的一个横截面面积不断减小的环管，环管上安装了一圈向中心伸出的支臂，用来安装喷嘴（见图2.5）。

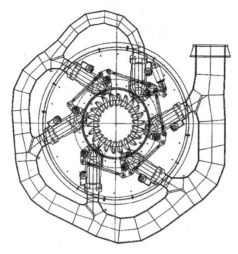

图2.5 六喷嘴切击式水轮机的配水管

2.7 机 壳

机壳的作用如下：
① 把水轮机转轮和多股射流包围起来，防止水外溅；
② 提供用来安装喷管的凹槽；
③ 提供安装轴承的凹槽；
④ 把机组的负荷传递给基础；
⑤ 作为基坑里衬；
⑥ 把水排入尾水渠。

由于水轮机在大气压下工作，所以机壳不起任何水力作用。

对于卧轴式机组（见图2.6），机壳由底部的一个矩形盒子和顶部的一个带有中间过渡段的半圆形罩组成。下面部分作为坚固的里衬保护周围的混凝土。下面部分是由焊接钢板制成的，配有加强筋和加强带以保证混凝土与机壳完美地附着在一起。同时还提供进检修门以方便检查。

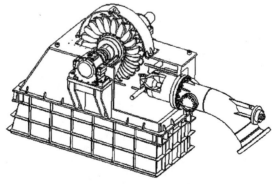

图2.6 切击式水轮机的机壳

在机壳内安装了一个铸钢盾板，在突然失控的状态下保护机壳不受射流的冲击。挡板用来破坏由于调速器的突然动作形成的偏转射流的能量。过渡段是指矩形盒子的顶部与水轮机中心轴之间的部分。过渡段上设有凹槽用来安装水轮机轴承。顶部的半圆形罩在主轴处被分为两半，拆下来后就可以看到转轮。半圆形罩的尺寸要精心设计以使转轮的风阻损失最小。

对于带有两个喷管的机组，在半圆形罩上设置一个倾斜的部分，这样不用拆除上部的喷管就可以取出转轮。

对于立式水轮机，机壳的形状可以是直径向下不断增加的锥形，或圆柱形，或顶部锥形、底部圆柱形。机壳是由钢板制造的。在机壳的外面安装了加强肋和加强钉，使机壳与周围的混凝土牢固地结合在一起。在侧边需要开孔使喷水管进入机壳。

2.8 喷射制动

即便冲击斗叶的所有射流都停止了，由于惯性转轮还是会继续旋转很长时间。为了在较短的时间内使转轮停止转动，设置了一个小喷嘴（副喷嘴），副喷嘴把射流导向斗叶的背面，使转轮反向旋转，如图 2.7 所示。该射流被称为制动射流。

副喷嘴直接从配水管中引水，需要时喷嘴会自动打开，当转轮停止运行时自动关闭，以避免转轮反向旋转。

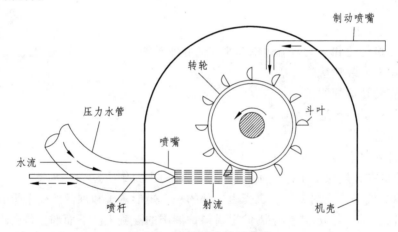

图 2.7 喷射制动的示意图

3 斜击式水轮机和双击式水轮机

3.1 斜击式水轮机

斜击式水轮机是一种适用于中水头的冲击式水轮机。斜击式水轮机是对切击式水轮机的一种改进，转轮为铸轮，形状就像外边缘封闭的风机叶片，如图3.1所示。斜击式水轮机的实际运行效率大约为87%，车间测试效率和实验室测试效率可以达到90%。这种水轮机的工作净水头在15米至300米之间。

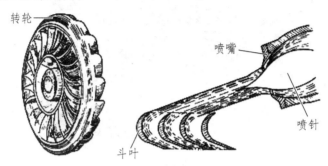

图 3.1 斜击式水轮机

3.1.1 优　点

Gilbert Gilkes 有限公司在1919年对切击式水轮机进行了改进，开发了斜击式水轮机。与混流式和切击式设计相比，斜击式水轮机在某些方面有一些优点。

第一，斜击式水轮机的转轮比切击式水轮机的转轮便宜。第二，斜击式水轮机与混流式水轮机不同，不需要密封的机壳。第三，斜击式水轮机的比转速高，与相同直径的切击式水轮机相比，工作流量更大，因此可以降低发电机成本和安装成本。除了比转速高、流量大、效率曲线平缓之外，斜击式水轮机还有结构简单、安装测试和调整简单、运行可靠、维护简单和成本低的特点。

斜击式水轮机的工作水头范围在混流式水轮机和切击式水轮机之间。尽管有很多大型的斜击式水轮机装置，但是斜击式水轮机在小水电中更流行，因为对这些电站来说降低成本是很重要的。

3.1.2 工作原理

斜击式水轮机是一种冲击式水轮机，在压力水管的末端设有一个特殊的喷嘴，可以把水流转换成高压射流。然后以与转轮面呈20度的角，把射流导向转轮叶片，叶片是勺形的，可以捕捉射流。图3.2为水射流冲击到斜击式水轮机转轮的一张照片。这些叶片的形状是精心设计的，压力水从叶片的一侧进入，流过叶片，然后从另一侧流出，把射流的动能转换为转动轴的功率。斜击式水轮机的转轮就像切击式水轮机的转轮从中间劈开似的。功率相同时，斜击式转轮的直径为切击式转轮的一半，因此比转速是切击式水轮机的两倍。

与其他冲击式水轮机不同，水流切向流过水轮机转轮意味着离开叶片的水不会影响下一个射流。在功率输入相同的情况下，斜击式水轮机可以采用较小的转轮直径和叶片直径以更高的转速驱动发电机。

斜击式水轮机的缺点是，为了产生足够的喷嘴压力使水轮机转轮高速转动，斜击式水轮机需要的水头比其他冲击式水轮机更高。

为了克服这个缺陷，可以在斜击式水轮机圆周上均匀地安装多个喷嘴，而不是仅采用一个喷嘴、一个射流来驱动转轮。在斜击式水轮机的转轮和叶片周围可以安装 2 个、4 个或 6 个射流喷嘴，从水流中获得更多的动能。

图 3.2　水射流冲击斜击式水轮机的转轮

射流喷嘴数量的增加意味着斜击式水轮机可以在低水头、低流量的条件下运转。由于功率输出与喷嘴数成正比，所以喷嘴数增加，水轮机从水中获得的能量也会增加。

与传统的水轮机和其他冲击式水轮机相比，斜击式水轮机改进了可用轴功率。斜击式水轮机的制造成本低，多喷嘴可以提高工作流量，所以一台较小的斜击式水轮机可以与一台较大的水轮机发出相同的电量。

3.2　双击式水轮机

双击式水轮机，又称班克式或米歇尔式水轮机、奥森博格式水轮机，是澳大利亚人安东尼·米歇尔、匈牙利人多纳特·班克和德国人弗里茨·奥森博格共同开发的。米歇尔在 1903 年获得了该水轮机的设计专利。匈牙利工程师班克教授对该水轮机进行了进一步开发。奥森博格把这种水轮机的制造引向批量生产阶段。其开发在 1933 年第一次被专利化。现在，奥森博格创建的公司是生产这种水轮机的主要厂家。

与大部分水轮机不同的是，双击式水轮机中水流是横向流过水轮机或水轮机叶片的，而不是轴向或径向的。水流先从水轮机外面流到水轮机里面，然后流过转轮，从另一侧流出，如图 3.3 所示。水流两次流过转轮，效率更高，但大部分能量都是在水流第一次进入转轮时进行转换的，水流离开转轮时，仅有 1/3 的能量转换。水流离开转轮时，顺便把转轮内的小碎片和污染物带走了。双击式水轮机是一种低转速机型，特别适用于低水头大流量的地方。

为了简化，图 3.3 中只可以看到一个喷嘴，实际上双击式水轮机都安装了两个喷嘴，布置时要避免水流相互干扰。

如果流量是变化的，可以把双击式水轮机设计成共用一个主轴但是流量不同的两个单元，如图 3.4 所示。两个单元中水轮机转轮直径相同，但是长度不同，用来处理相同压力不同流量的水。两个转轮划分的体积比通常为 1 : 2。较窄的单元处理较小的流量，较宽的单元处理中等流量。两个单元一起处理全部流量。分段的调节机构，即水轮机上游段的导叶系统，可以根据流量进行调节、灵活运行，出力可以为33%、66%或100%。由于水轮机构造相对简单，所以运行成本较低。

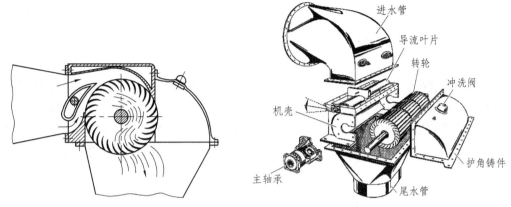

图 3.3 双击式水轮机的示意图　　　　图 3.4 奥森博格式水轮机断面图

3.2.1 效 率

双击式水轮机的最高效率略小于轴流转桨式、混流式或切击式水轮机。低水头的小型双击式水轮机的总效率在整个流量范围内为80%~84%。高水头的大中型双击式水轮机的最大效率可达87%。但是，双击式水轮机的效率曲线在负荷变化时很平缓。采用拼合式转轮和拼合式水轮机室，当流量与负荷在1/6到最大之间变化时，双击式水轮机的效率均可以保持不变。可以用效率曲线来说明双击式水轮机在部分负荷时的优点，如图3.5所示。

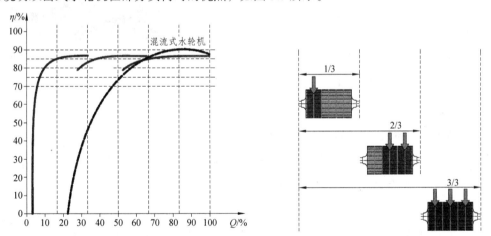

图 3.5 导叶按照 1∶2 的比例进行流量调整时双击式水轮机的效率曲线与
混流式水轮机的效率曲线

特别是对于小型河床径流式电站，效率曲线平缓的水轮机的年度性能比其他水轮机系统更高，原因为小型河流在某些月份流量很少。水轮机的效率决定了在河流流量较小的时段是否可以发电。如果采用的水轮机最高效率很高但部分负荷时效率很低，则年度性能将低于效率曲线平缓的水轮机。

3.2.2 水轮机机壳

双击式水轮机的机壳是由结构钢制成的，具有结实、耐冲击和抗霜的特点。如果水中含有大量的磨蚀物质（如沙或淤泥）或水体的实际成分具有腐蚀性（如海水或酸性水），水轮机与水体接触的所有部件都应由不锈钢制成。

3.2.3 导 叶

在拼合式双击式水轮机中，两个受力平衡的异形导叶引导工作水流。导叶把水流分为两部分，引导水流平滑地进入不同开度下的转轮。两个旋转导叶都被精准地安装在水轮机壳中，水头低时可以作为水轮机的关闭装置。因此导叶可以作为压力水管与水轮机之间的阀门，不再需要使用截流阀。通过调整杆自动或手动控制可以对两个导叶进行单独调节。导叶安装在高度抗滑的轴承中，滑动轴承不需要任何维护。停机时，通过臂端增加的重量，导叶可以靠重力自动关闭。

3.2.4 转 轮

转轮是水轮机中最重要的组成部分。双击式水轮机的转轮是圆柱形的并配有一个卧轴。转轮上安有叶片，叶片是由精轧的异形钢制成的。为了降低对水流的阻力，把叶片的边缘削尖。叶片的横截面是圆弧形的（就像沿着长度方向切削后的管子）。叶片的末端焊接到圆盘上形成一个类似于仓鼠笼形状的笼子，因此有时这种水轮机又被称为"鼠笼式水轮机"；区别在于双击式水轮机采用的是槽形的钢制叶片，而"鼠笼式水轮机"是条状的，如图3.6所示。转轮上的叶片数最大为37片，取决于转轮的尺寸。线性倾斜的叶片只产生很少的轴向力，因此不再需要带有复杂配件和润滑装置的加固止推轴承。对于宽度较大的叶轮，可以采用多个圆盘对叶片进行支撑。

图3.6 双击式水轮机转轮示意图

3.2.5 轴 承

双击式水轮机配有调心滚子轴承，该轴承的优点为滚动阻力低、维护简单。轴承箱在设计时要防止水渗入轴承，防止润滑油与工作水流相接触。此外，轴承把转轮固定在水轮机壳的中心。除了每年要更换润滑油之外，不需要对轴承进行任何其他维护。

3.2.6 尾水管

与切击式水轮机类似，双击式水轮机是一种自由射流的水轮机。然而在中低水头时，为了利用全部的水头，可以安装一个尾水管。但尾水管中的水柱必须是可控的。调节空气阀可以调节水轮机机壳的吸入压力，进而控制水柱高度。这样可以对吸入压头为1至3米的水轮机进行优化，避免发生空化。

3.2.7 优 点

由于成本低、调节性能好，双击式水轮机广泛应用于功率小于2000千瓦、水头低于200米的小型和微型水电机组中。

由于部分负荷时性能很好，双击式水轮机非常适合于不需要值班的发电厂。双击式水轮机的结构简单，维护起来比其他水轮机更容易；该水轮机仅有两个轴承需要维护，且仅有 3 个转动部件。该水轮机的机械系统很简单，可以由当地的机修工进行修理。

另外一个优点是，双击式水轮机可以自我清理。随着水流流出转轮，叶子、杂草等将被带出转轮，减少损失。因此，尽管水轮机效率稍低，但比其他水轮机更可靠。通常不需要清理转轮，可以通过反转水流或改变转速进行自我清理。其他类型的水轮机更容易堵塞，虽然额定效率很高，但是功率损失也比较大。

4 混流式水轮机

4.1 简 介

法兰西斯式水轮机是根据美国工程师詹姆斯·比切诺·法兰西斯来命名的,法兰西斯对1849年就存在的反击式水轮机进行了改进。

为提高效率,不断地对最初的法兰西斯式水轮机进行改进,现在法兰西斯式水轮机的效率超过了93%。

这种水轮机使用范围很广,工作水头在25米至550米之间。

法兰西斯式水轮机是一种反击式水轮机,高压水进入水轮机转轮,水流流过弯曲的叶片产生反作用力,把压能转换为机械能。由于水轮机内压力是变化的,这种水轮机也被称为变压水轮机。由于水流径向进入转轮,向内流动,然后流向改为轴向,所以法兰西斯式水轮机又被称为混流式水轮机。

4.2 主要组成部件

混流式水轮机的主要组成部分为:蜗壳、固定导叶、活动导叶、转轮和尾水管,如图4.1所示。

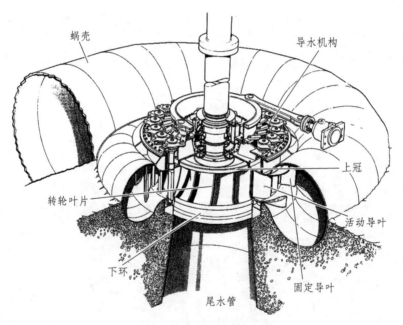

图 4.1 混流式水轮机

4.2.1 蜗 壳

蜗壳是构成水轮机外周的环状螺旋形机壳。蜗壳的一端与压力水管相连,接收压力水,并把水均匀地分配到整个转轮的圆周上。蜗壳的横截面面积是递减的,这样可以保证更多的水导向转轮时,流速保持恒定。蜗壳的内表面是圆柱形的,固定导叶环就安装在内表面上。

4.2.2 固定导叶环

固定导叶环（座环）具有环状结构，上下两个圆环之间焊有很多鳍形的固定导叶，如图 4.2 所示。固定导叶的进口尺寸与蜗壳内表面开口的尺寸相同。固定导叶又称为固定的导流叶片。

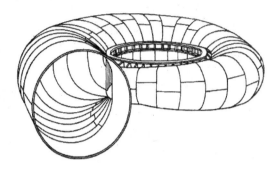

图 4.2 带有固定导叶环的蜗壳

固定导叶的作用如下：

① 把从蜗壳引来的水以适当的角度导向活动导叶，因此必须精心设计固定导叶的形状。

② 对蜗壳进行加固以应对流过固定导叶的高压水流。因此，固定导叶环是牢固地焊接在蜗壳上面的。安装完成后，蜗壳和固定导叶周围都灌满混凝土。

③ 把发电机主推力轴承的负载传递给基础。

④ 为顶盖提供连接装置。

4.2.3 导水机构

导水机构（见图 4.3）有 3 个功能。

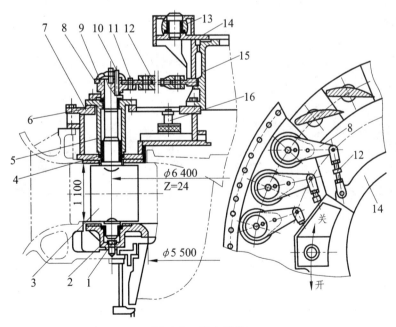

图 4.3 导水机构

1、4、6—尼龙轴瓦；2—底环；3—导叶；5—轴套；7—顶盖；8—连接板；9—转臂；10—分半键；
11—剪断销；12—连杆；13—推拉杆；14—控制环；15—支座；16—补气阀

① 接收来自固定导叶的水，并以适当的角度导向转轮叶片，把转轮进口的冲力降为最小。

② 通过改变进水量来调节水轮机，适应瞬时负荷的变化。使活动导叶绕各自的轴旋转，改变两个叶片之间通道的横截面面积。

③ 作为一个阀门在关闭时截断水流。

活动导叶位于固定导叶和转轮之间。叶片是翼形的，这样水流流过叶片后不会分流。在关闭的位置，任何活动导叶的首端必须与下一个导叶的尾端紧密地连在一起，以免漏水。每个活动导叶都有自己的轴，可以绕轴旋转。活动导叶轴的下轴颈安装在轴套内，轴套与套筒配合安装于底环内。活动导叶轴的上轴颈通过导轴承穿出顶盖。连杆和转臂把所有这些突出的轴连接到一个叫控制环的圆环上。顺时针或逆时针稍微转动控制环就可以同时旋转所有活动导叶，从而打开或关闭导叶。控制环由两个活塞接力器带动，这两个活塞接力器与控制环相切，间距180度。调节机构以同样的幅度把一个活塞向前移动，另一个活塞向后移动（类似于双手驾驶汽车方向盘的动作）。

活动导叶的数量比转轮叶片数要多很多。通常有18至26个活动导叶，但是某些情况下可以达到32个。

4.2.4 转 轮

转轮是水轮机的核心部件。转轮位于活动导叶和尾水管之间。水流从转轮整个圆周进入转轮，流过转轮叶片时把大部分能量传递给叶片，然后流入尾水管。

水流径向向内进入转轮，然后逐渐改变方向，流到转轮出口时变为轴向。转轮主要由三部分组成：上冠、叶片、下环，如图4.4所示。为了最大限度地进行能量转换，必须精心设计转轮叶片的形状。必须把水力摩擦和涡流损失降到最小。叶片上边缘与上冠相接处呈严格的流线型。上冠的外表面上面开孔用来与转轮轴相连。如果转轮进口直径比出口直径大，则下环表面也要呈流线型。否则，下环呈圆环状。

图4.4 低水头（左）和高水头（右）混流式水轮机的转轮

转轮的形状取决于水头、出力和水轮机的比转速。

转轮的叶片数取决于水头、出力和转轮的转速，在10至20之间。叶片数可能会小于活动导叶的数量。

为了防止压力水绕过转轮，直接从活动导叶流到尾水管，在转轮出口处下环的外表面安装被称为迷宫环的密封圈，如图4.5所示。

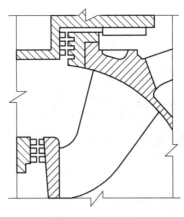

图 4.5 迷宫环

4.2.5 尾水管

尾水管位于转轮的外侧。尾水管的一端与转轮出口相连,另外一端与尾水渠相连。尾水管的功能如下:

① 把从转轮流出的水排到尾水渠中。

② 在转轮出口形成负压,提高转轮的有效水头。安装尾水管后,作用在转轮叶片上的总水头为静压头(等于上游水位与转轮出口截面的高程之差)与转轮下的负压头(真空)之和。这样就可以把转轮安装在尾水渠以上,并减少昂贵的基础工程的施工。

③ 把速度能转化为压能,从而回收速度能,增加水轮机的效率。转轮出口处的速度头可以达到有效水头的 10%~30%。必须精心设计尾水管的形状以达到最优的状态。

最常见的尾水管是肘形的,锥管段后面是弯管,然后是具有椭圆形或矩形断面的扩散段,如图 4.6 所示。

出口面积与进口面积的比值随着回收率的增加而增大。如果出口面积太大,可以在水流方向把扩散段用隔墙分为几个部分。

低水头的水轮机,水轮机的出口直径比进口直径大。可以从下面取出转轮进行检查和维修,这时需要在尾水管上安装一个可以拆卸的管段。

尾水管里衬安装好后,在其周围要灌满混凝土,还需要配备必要的紧固件、肋板等。通常,尾水管的出口段本身就可以用混凝土进行浇筑。另外还需要设置一个检查口,以随时进入尾水管进行检查。

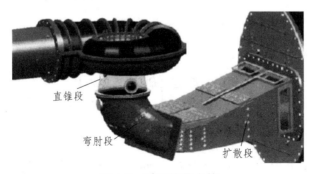

图 4.6 肘形的尾水管

4.2.6 顶 盖

顶盖是位于转轮上表面的一个圆柱形箱式结构,如图 4.7 所示。

图 4.7 混流式水轮机的顶盖

顶盖的功能有:① 防止水通过活动导叶与转轮之间的间隙流到水轮机外面;② 在周围安装螺栓与水轮机主体相连;③ 提供圆形开孔安装导轴承,方便导叶耳轴伸出顶盖;④ 在中心开孔安装水轮机主轴和填料盒的主导轴承,开孔的尺寸必须大于主轴法兰的直径;⑤提供与控制环和其他附属机构的连接。

4.2.7 底 环

底环位于固定导叶环和尾水管之间、水轮机活动导叶的底部,如图 4.8 所示。底环包括上孔环和下孔环两部分,上孔环的外径与固定导叶的内径相同,下孔环的直径约等于尾水管的直径。下孔环上开了一些孔,使活动导叶耳轴的下轴颈可以穿入轴承中。

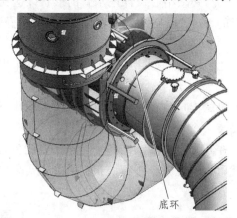

图 4.8 混流式水轮机的底环

4.2.8 主 轴

大部分现代的混流式水轮机采用的都是立轴,这样发电机可以放在最高尾水位高程以上,避免淹没发电机,根据设计参数,水轮机可以安装在另一高度。厂房内发电机和水轮机放在不同的楼层,安装工作可以单独进行。

可以直接用螺栓把水轮机主轴与发电机主轴相连。如果主轴太长,可以设置一个中间轴,这样需要取出水轮机转轮时,只需拆掉中间轴而不会影响到发电机轴。

水轮机和发电机组成的整个立式液压装置仅有一个推导组合轴承。该轴承一般位于发电机转子的下面，有时位于水轮机机壳的上面。机组中有两个导轴承，一个在发电机轴的上面，另一个在水轮机轴的上面。

目前，抽水蓄能发电站中的水电机组采用的是卧轴混流式水轮机，每个机组包括一台水轮机、一台发电电动机和一台水泵。这种布置方式很少纯粹用来发电。转轮是悬挂安装在主轴上的。如果需要为转轮配备轴承，主轴需要穿过尾水管。这样设计是有利的，特别是在部分负荷运行时，因为可以避免在弯管处产生真空。

4.2.9 机坑里衬

机坑里衬是一个圆柱形的里衬，里衬直径很大，可以把水轮机顶盖作为整体一起取出，里衬高度从固定导叶环的顶部开始一直延伸到水轮机层，如图 4.9 所示。在里衬的外侧浇筑混凝土，里衬作为骨架。在里衬的外表面安装必要的肋板、螺栓及其他紧固件。

图 4.9 蜗壳上已经安装好的基坑里衬

4.2.10 水轮机排水

对于小型水轮机，为了方便排水检修，在上游面设置一个阀门，在尾水管末端设置一个尾水管闸门。把阀门和闸门关闭后，就可以把水轮机的空间放空了。对于大型水轮机，不能设置阀门，所以在压力水管进口处设置一个闸门以控制水流。

5 轴流定桨和轴流转桨式水轮机

定桨式和转桨式水轮机是反击式水轮机中的轴流式水轮机。定桨式和转桨式水轮机的区别在于定桨式水轮机的转轮叶片是固定的，而转桨式水轮机的叶片是可以调整的。

5.1 开 发

5.1.1 定桨式水轮机的开发

在大多数电站的厂址处，大量的水只在低水头时能够利用，此时混流式水轮机的尺寸将变得很大而不经济，为此开发了另外一种水轮机——轴流定桨式水轮机。

这种水轮机的主要部件，如蜗壳、固定导叶、活动导叶和尾水管，都与混流式水轮机类似，如图5.1所示。但是转轮在以下方面有所不同：

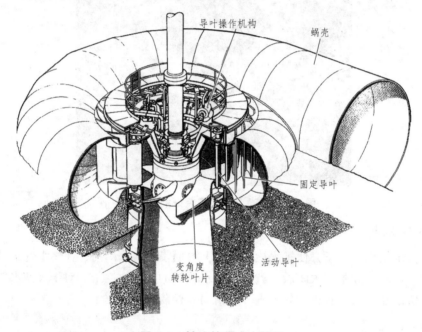

图 5.1 轴流转桨式水轮机

① 轴流定桨式水轮机的转轮是轴流式的，水流平行于主轴进入和流出转轮，而混流式水轮机的转轮是混流的。

② 转轮叶片类似于螺旋桨的叶片。叶片是安装在转轮轮毂上的悬臂结构，叶片的外缘并没有连接在一起（转轮没有下环）。

③ 叶片的数量在4至8之间，比混流式水轮机少得多。因此，叶片所占的空间也少，过流面积更大。

轴流定桨式水轮机在额定出力及邻近区间时效率很高，但在部分负荷时效率较低。因此，这种水轮机只能在出力恒定的电站中使用。

5.1.2 转桨式水轮机的开发

很多人都试图改变轴流定桨式水轮机在部分负荷时的效率，卡普兰成功地设计了一种转轮叶片可以旋转的轴流式水轮机，可以改变叶片的倾角以适应瞬时负荷的变化。轴流转桨式水轮机是根据维克多·卡普兰（1876—1924）教授的名字命名的。该水轮机的工作水头范围为5米至60米，特殊情况下可达到80米。该水轮机比转速为300～1000。

转轮叶片的倾角要不断地改变，因此必须提供一个与导叶操作机构类似的调节机构。因为转轮叶片每时每刻都在旋转，所以要求调节机构也要跟着随时转动。

5.2 转轮的主要组成部分

转轮的主要组成部分有：① 轮毂；② 带操作机构的叶片；③ 泄水锥，如图5.2所示。

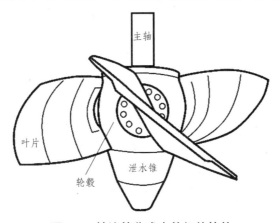

图 5.2 轴流转桨式水轮机的转轮

5.2.1 轮 毂

轮毂是带有球形中腹部的中空圆柱体。转轮叶片安装在轮毂周围专门设置的凹槽内，耳轴伸入轮毂体中。操作机构设在轮毂体内。轮毂内充满了润滑油，为运动部件提供可靠的润滑。轮毂体内的油压大于外面的水压以防止水渗入到润滑油中。

泄水锥位于尾水管侧的轮毂末端。根据设计，泄水锥可以是一个整体，也可以由两部分组成。在设计时，应保证泄水锥对水流的阻力最小。在轮毂的周围设置必要的紧固件与泄水锥相连。轮毂的另一端与水轮机主轴相连。

5.2.2 叶 片

叶片由以下部分组成：接收水流的翼型悬臂部分，看起来像圆环的一个弧；支撑翼型悬臂部分的耳轴，把产生的水推力和离心力传递给轮毂。

叶片必须足够宽，即使在全开的位置，相邻叶片也要充分地相互重叠，避免水通过两个叶片间的空隙直接流到尾水管中。

叶片数以及转轮轮毂直径随着水头增加而增加。四个叶片的适用水头为20米，而八个叶片的适用水头在40米至80米之间。

可以用螺栓把叶片固定到耳轴上，然后把耳轴安装到转臂上，形成控制机构的一部分。在轮毂凹槽中提供特殊的填料，防止水渗入轮毂体，也防止轮毂体内压力大于最高水头时，轮毂体内的润滑油向外渗透。如果由于意外，水渗入转轮轮毂内，水会沉积到轮毂的底部，通过专门设计的出口靠空气压力把水排出。

通过转轮叶片操作机构，可以使叶片的倾角在 0 度至 30 度之间变化，所需的操作时间为 15 至 30 秒。

5.2.3 转轮叶片操作机构

叶片操作机构的作用是改变叶片的角度以配合活动导叶操作机构满足瞬时负荷的变化。叶片操作机构由以下部分组成：接力器、主叶片操作杆、十字头、曲柄连杆、耳轴、压力受油器，如图 5.3 所示。

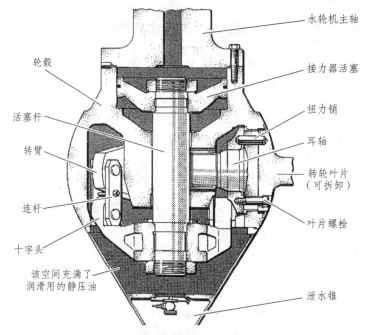

图 5.3 转轮叶片操作机构

接力器是一个缸体，在缸体内通过油压使接力器活塞轴向运动。叶片操作杆与活塞刚性连接，带动曲柄连杆和转臂动作，改变转轮叶片的角度。轴流转桨式水轮机还配备了位于发电机或励磁机上面的受油器以及位于发电机和水轮机主轴内部的立式旋转油管。受油器把压力油从固定的调速器油管输送到旋转油管，再输送到转轮叶片接力器中。

图 5.4 中给出了转轮叶片斜率调整系统的一个实例。通过活塞（4a）的活塞杆（5）带动转轮叶片做旋转运动，进而调整转轮叶片（1）的斜率。水轮机主轴（6）上端延伸出的圆柱体作为接力器缸（4），而发电机轴（8）下的法兰作为缸盖。活塞杆（5）在两个轴承（7）之间运动。

接力器（4）的供油装置位于发电机轴（8）的上端。通过中空的发电机轴内布置的两个同轴管（9 和 10），把油输送到接力器的相应位置。在内管（9）中，润滑油往返于活塞（4a）下侧与润滑油室（12a）之间，在同轴管（9）和（10）之间的环形空间内，润滑油往返于活塞（4a）下侧与润滑油室（12b）之间。注油口（12）位于机组上面，设有两个润滑油室（12a）和（12b）用来提供润滑油。

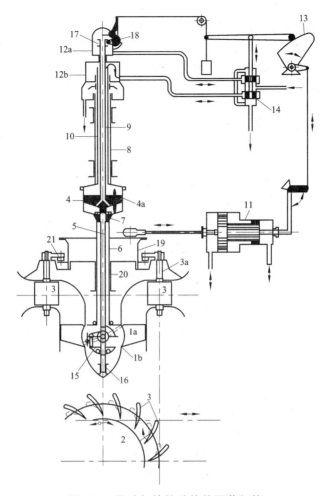

图 5.4 导叶与转轮叶片的调节机构

1—转轮叶片；2—导叶叶栅；3—导叶；4—转轮叶片接力器；4a—接力器活塞；5—接力器活塞 4a 的传力杆；
6—水轮机轴；7—轴承；8—发电机轴；9、10—同轴管；11—导叶栅接力器；12—注油口；
12a、12b—润滑油室；13—凸轮；14—使高压油回到转轮接力器的调节阀；
15—活塞杆 5 的传力转臂支架；16—活塞杆 5 的传送；17—反馈杆内衬；
18—反馈杆；19—控制环；20—径向轴承；21—导叶臂

5.2.4 导叶调节与转轮叶片调节的配合

为了在整个负荷范围内获得最佳的效率，转轮叶片的角度与活动导叶的角度之间是相互关联的。两者之间的关系是水头的函数。在水头和负荷变化时，调速装置可以自动调整两个接力器的动作以达到最优效果。

水轮机调速器直接通过接力器（11）控制导叶（3）的动作，如图 5.4 所示。接力器的动作带动并控制转轮叶片斜率的调整。接力器（11）通过连杆和转臂把动作传递给凸轮（13），而凸轮的动作又受到接力器活塞（11）的影响。用这种方法，柱形阀（14）移动到中间位置，通过供油压力使接力器活塞（4a）动作。

5.2.5 接力器的位置

接力器可以布置在 3 个位置：

197

① 在转轮轮毂内，如瑞典的拉萨尔电站。这种布置方式，活塞杆很短，但是水头高时，接力器会变得很大、很笨重。

② 在水轮机轴和发电机轴之间。这种布置方式会增加机组的尺寸。

③ 在发电机轴的上面，如巴西的特雷斯马里亚斯电站。这种布置方式，水电机组很紧凑，但是活塞杆很长。

5.3 蜗　壳

压力水管中的水进入蜗壳，流向转轮周围，如图 5.5 所示。蜗壳使水流在水轮机转轮的圆周上均匀分配，保证配水时流速恒定。

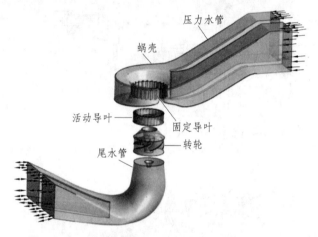

图 5.5　轴流转桨式水轮机的总体布置

水头小于 30 米时，蜗壳可以用混凝土进行浇筑，可以采用梯形截面。

水头较高时，水压力可能太大，超过混凝土可以承受的负载。在这种情况下，采用钢板制作蜗壳，与混流式水轮机的蜗壳类似。蜗壳的横截面通常是圆环形的，钢板壳焊接到座环上。

即使是混凝土蜗壳，也需要安装座环。座环上的叶片把水从蜗壳引向活动导叶。此外，座环和固定导叶还要传递液压动力，因此采用大型的预应力拉杆螺栓把座环固定在混凝土中。座环通常由钢板焊接而成，内部填满混凝土。

5.4 活动导叶栅

轴流转桨式水轮机的活动导叶栅与混流式水轮机相似，但轴流式水轮机的叶片更高，相邻叶片之间的中心距离更短。运行时，水轮机负荷发生变化，控制环以同样的角度同时转动导叶。活动导叶是由钢板材料制成的，耳轴焊接在导叶上。叶片设计要确保水力流动条件最优，还要保证表面的光滑度。

5.5 顶盖和尾水管

随着水轮机物理尺寸的增加，顶盖被分为两部分。外面部分钻孔使活动导叶耳轴穿出。里面部分呈中空双曲面状，如图 5.6 所示。

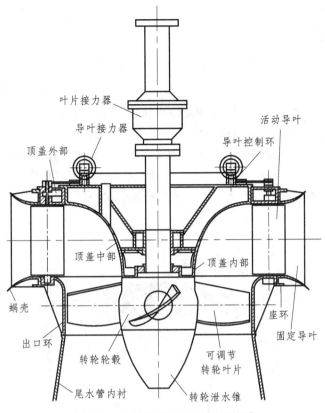

图 5.6 轴流转桨式水轮机的构造

尾水管是截面面积不断增加的一个管道，用来把从水轮机流出的水排到尾水渠中。尾水管把水轮机出口废弃的动能转换为有用的压能。

轴流转桨式水轮机的尾水管相对更大。转轮出口处的速度头可以达到有效水头的40%。由于尾水管要回收大部分的水头，其设计很重要。出口处的横截面面积很大，可以分为两部分或三部分。尾水管是肘形的，可以完全用混凝土进行浇筑，进口部分要增加钢板内衬。

5.6 自动排气阀

在顶盖上面安装自动排气阀，甩负荷时突然关闭活动导叶后，用来破坏转轮顶部可能形成的真空。

5.7 水电机组的主轴

轴流式水轮机的主轴是中空的，用来容纳转轮叶片操作机构。

5.8 超速保护装置

如果水轮机的转速超过同步转速的30%，超速保护开关将关闭机组，同时全部打开转轮叶片以降低转速。

6 斜流式水轮机

6.1 简　介

德里亚水轮机又被称为斜流式水轮机。该水轮机是以其发明者液压设计工程师保罗·德里亚命名的。在本质上，斜流式水轮机就是一种改进的轴流转桨式水轮机，更适合安装较多的转轮叶片。这种水轮机的流道向主轴倾斜，转轮的叶片可以转动，如图 6.1 所示。斜流式水轮机的适用水头达 200 米，效率超过 92%。斜流式水轮机还可以用作可逆式水泵水轮机，提升扬程达 60 米。已经证实，斜流式设计适用于较高的水头以及一些抽水蓄能电站。

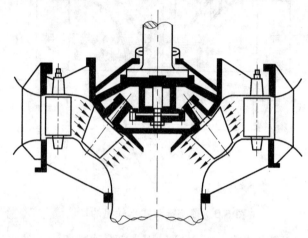

图 6.1　斜流式水轮机示意图

斜流式水轮机是一种双重调节的反击混流式水轮机。自 1999 年以来，Geppert 公司就已经开始为小型的水电站制造这种新型水轮机，并成功安装了几十个。1958 年，在苏格兰卡利格兰的地下电站中安装了第一台不可逆的斜流式水轮机，其水头为 55 米，输出功率为 22750 千瓦。在尼亚加拉大瀑布的加拿大电站和美国电站中都安装了早期的斜流式水轮机系统。

可调节的转轮叶片和导叶使这种中压水轮机在很大的流量范围内都可以保持较高的效率，与轴流转桨式水轮机相比，这种水轮机在部分负荷时效率更高。斜流式水轮机的效率略低于混流式水轮机的最高效率，但混流式水轮机仅在设计水头与负荷范围内效率才高。相反，斜流式水轮机的叶片可以调整以适应水头和负荷的变化，因此斜流式水轮机的总体效率高于混流式水轮机。对于流量季节性变化很大的电站，斜流式水轮机的年发电量要高于可比较的混流式水轮机的发电量。

当季节性流量变化很大时，通常采用斜流式水轮机，以避免采用多台混流式水轮机机组的昂贵方案。由于斜流式水轮机对空化的敏感性较弱，安装位置较高的轴流转桨式水轮机都被斜流式水轮机取代了。

6.2 开　发

第二次世界大战后，水力发电对适用于 60 米以上高水头轴流转桨式水轮机的需求越来越大。但是轴流转桨式水轮机的水头增加后，会出现以下问题：
① 轮毂直径增加，转轮叶片的径向尺寸减小；
② 叶片数增加，同时连杆和转臂数也要增加。

轮毂内中空部分的空间不能容纳更大的叶片操作机构。此时传统轴向布置的轴流转桨式水轮机的转轮轮毂将变得非常大，叶片将变得非常窄，以至于水轮机的效率明显降低。按照逻辑，应该把轴流式改回混流式水轮机的混流式布置方式，结果就形成了叶片可以转动的混流式水轮机，如图6.2所示。

图6.2 斜流式水轮机转轮

这种水轮机的蜗壳、固定导叶、活动导叶和尾水管与轴流转桨式水轮机完全相同，不同的是导叶与尾水管之间的流道向主轴倾斜。转轮叶片的轴也不再与主轴垂直，而是向主轴倾斜，与流道垂直，这样布置可以克服高水头时对轴流转桨式水轮机的水力限制。与轴流转桨式水轮机的轴向流不同，此时水流的流向为斜向（对角的）。

斜流式水轮机倾斜的叶片使其特别适用于20米和100米之间的水头范围——这正是轴流转桨式水轮机和混流式水轮机交界的工作范围。由于转轮叶片是可调的，斜流式设计还有很多其他优点：在很宽的水头和负荷范围内都可以平稳高效的运行；压力和负荷在整个叶片上均匀分布（即从蜗壳到轮毂的中跨处）；在整个运行范围内都不会产生空化；比转速和飞逸转速较低；水推力较小。

6.3 机械特性

在轴流式水轮机中，转轮轮毂是圆柱形的，叶片轴垂直于转轮轴。斜流式水轮机的一个特色就是叶片的轴线与水轮机旋转轴之间的夹角为锐角，锥形的转轮轮毂不会限制水流，因此可以增加叶片的数量，并可以应用于压力较高的工况。在斜流式水轮机中，水流相对于水轮机叶片既不是轴向也不是径向，而是呈一定角度。叶片轴线与主轴之间的夹角θ随着水头的增加而减小，在30°至60°之间变化，如图6.3所示。

转轮轮毂的形状是锥形的，与轴流转桨式水轮机相比，轮毂更矮。但是斜流式水轮机轮毂的基座更大，有足够的空间可以容纳叶片增加后的转臂和连杆机构。

接力器的设计更加简单。

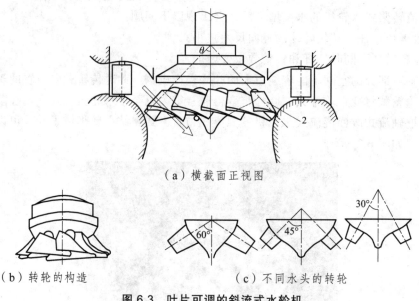

（a）横截面正视图

（b）转轮的构造　　　　（c）不同水头的转轮

图6.3　叶片可调的斜流式水轮机

1—转轮轮毂；2—叶片

6.4　水力特性

由于进口面积较大，所以流速较低。由于进口线速度大于出口线速度，可以把叶轮的进口角与出口角做成相同的。这样可以提高部分负荷时的效率，减少空化的倾向。

在轴流转桨式水轮机中，水流要先转一个90度弯变为轴向后再进入转轮室，这会影响水的流速，减小水的能量。与轴流转桨式水轮机不同，斜流式水轮机倾斜的流道使水流流态更加直接，明显的变化较少。转轮的表面积更小，这两点都可以降低摩擦损失。

图6.4对比了斜流式水轮机与径流-轴流式水轮机的特性。图中η/η_{max}为运行效率与最大效率的比值；N/N_{opt}为运行功率与最优功率的比值。运行时的负荷和水头与计算值差别较大，但是由于水流流过轮毂叶片和吸水管时比较平滑，斜流式水轮机的流动状态更加平稳，脉动更小，效率曲线更加平缓，平均运行效率η更高。斜流式水轮机的空化特性稍微差于径流-轴流式水轮机。因此，斜流式水轮机可以安装在水头达200米的水电站中，取代了该范围内径流-轴流式水轮机的使用。在压头和功率波动很大的电站中，斜流式水轮机更加经济实用。

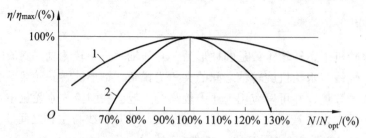

图6.4　斜流式（1）和径流-轴流式（2）水轮机的特性比较

6.5　失控（飞逸）工况

斜流式水轮机在超速时流量会降低。如果水轮机开始空转并向飞逸转速逼近，累积起来的离心水头将使流量减小。这种抑制机制随着水轮机转速的提高而增大。与轴流转桨式水轮机相比，这种自我调节的机制可以降低斜流式水轮机的飞逸转速（突然失去负荷），因此，采用的发电机可以更小一些，从而大大节约了发电机的成本。同时还可以减小桥式吊车的尺寸，节约更多成本。

6.6　可逆性

斜流式水轮机的转轮还被广泛用于抽水蓄能电站内可逆式水力机械（水泵水轮机）中。斜流式水轮机可以作为水泵使用，扬程可达 90 米，而轴流转桨式水轮机作为水泵时的最大扬程还不到 20 米。20 世纪 50 年代末，在加拿大安大略省水力发电委员会的亚当贝克爵士 2# 抽水蓄能电站中首次安装了六台可逆斜流式水泵水轮机。

6.7　接力器

轴流转桨式水轮机中转轮叶片操作机构的设计非常繁琐。对于水头较高的水轮机，必须把叶片操作机构放在转轮外面，在水轮机轴和发电机轴之间或在发电机上面，需要很长很重的活塞缸。

在斜流式水轮机中，调节叶片角度的机构位于轮毂内，可以采用简单的旋转轴式接力器机构，如图 6.5 所示。

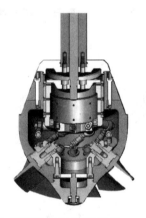

图 6.5　斜流式水轮机的转轮和八个倾斜叶片的调节系统

通过活塞杆把接力器活塞的旋转动作传递给转轮轮毂内的三脚架。三脚架与与叶片臂通过滑块相连。滑块把三脚架的旋转运动传递给转轮叶片臂，进而改变叶片的攻角，适应水头和负荷的变化。

7 灯泡贯流式水轮机

7.1 简 介

灯泡贯流式水轮机是一种轴流转桨反击式水轮机,适用于超低水头。灯泡贯流式水轮机的特点是水轮机的主要组成部分及发电机都位于灯泡体内,这也是其名称的来源。与轴流转桨式水轮机的主要区别在于水流进入导叶叶栅时水流的方向为混流(轴向-径向流),而且没有蜗壳。导叶轴与水轮机主轴呈一定角度,一般为 60°。与其他类型的水轮机不同,灯泡贯流式水轮机的导叶叶栅是锥形的。

灯泡贯流式水轮机的转轮与轴流转桨式水轮机的转轮采用的设计方式相同,根据水头和流量的不同,可以选择不同的叶片数。

7.2 总体布置

图 7.1 为灯泡贯流式水轮机电站总体布置的纵剖面图。图 7.2 给出了详细的水轮机设计图。

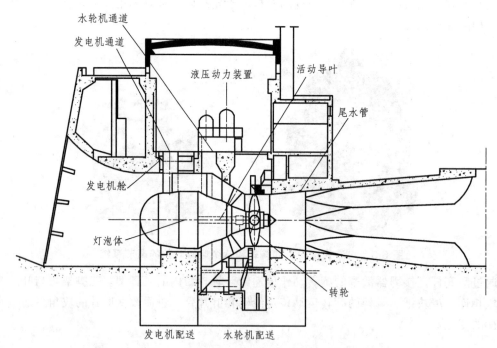

图 7.1 灯泡贯流式水轮机的总体布置图

水流沿着轴线流向位于管道中心的机组,流过发电机、主支柱、活动导叶、转轮和尾水管后,进入尾水渠。

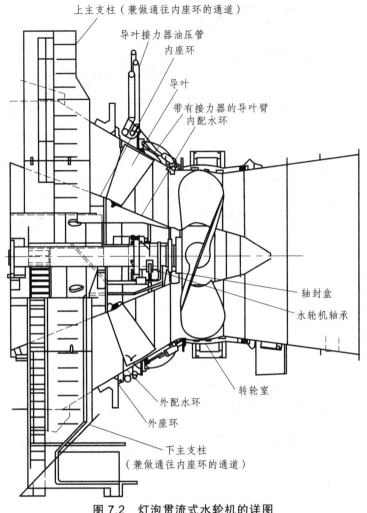

图 7.2 灯泡贯流式水轮机的详图

7.3 主要组成部分

灯泡贯流式水轮机主要由以下部分组成：座环、转轮室、尾水锥、发电机舱、导流板、旋转部件、水轮机轴承、轴封盒和导叶操作机构。

7.3.1 座环

图 7.3 为座环构造和尾水管的纵断面图。图中（1）为下主支柱，（5）为上主支柱。内座环（3）和外座环（6）都焊接在主支柱上。用螺栓把内配水环紧固在内座环的下游侧，如图 7.2 所示。

外座环构成水管外壁的一部分，与主支柱的外部组件一起预埋在混凝土中。发电机灯泡体被固定在内座环的上游侧，如图 7.1 所示。这些部件都位于水流中，与转轮轮毂一起构成水管内壁。

在主支柱的上游灯泡体两侧还安装了两个侧支柱（4）和（7），用来对灯泡体进行加固，防止产生共振。

座环上的两个主固定导叶构造把总重量和液压力传递给周围的混凝土。通过该构造把水轮机和发电机的动态力和静态力传递给建筑基础。

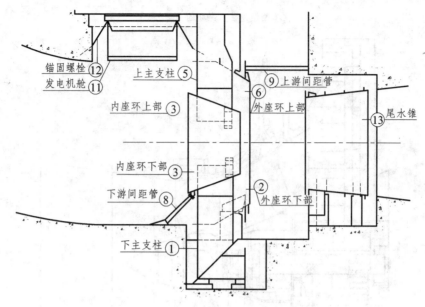

图 7.3 座环和尾水管的纵断面图

7.3.2 转轮室与尾水锥

转轮室是外座环与尾水管的连接部分,如图 7.2 所示。外座环下游面的法兰用来连接转轮室。

尾水锥由两个或多个直的钢制锥体焊接而成,并预埋在混凝土中。上游面与转轮室通过柔韧的伸缩节相连。伸缩节允许转轮室和外配水环进行一定的轴向运动,以适应由于温度变化引起的拉伸或收缩。

钢锥内衬的长度取决于出口处的最大流速,并且应保证不能破坏混凝土。

7.3.3 发电机舱

图 7.4 中所示的发电机舱(11)通常与水轮机一起配送。发电机舱位于发电机的上面,并提供检修门进行发电机的安装或拆卸工作。

发电机舱由一个穿孔部分组成,构成发电机舱开口处的水管外壁。舱盖由一个上面带有法兰的圆柱形盖板构成,同时用来安装密封装置。由于机组的灯泡体随着水轮机的充水和排水而上升和下降,盖板和舱盖之间的密封接头必须允许盖板进行垂向运动。

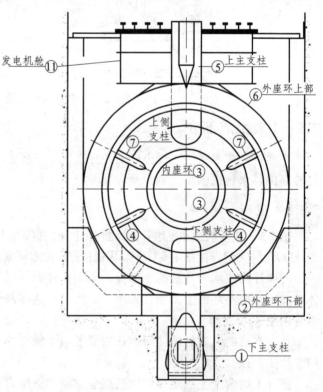

图 7.4 座环的横截面图

7.3.4 导流板

导流板位于发电机进口竖井和水轮机主支柱之间。导流板为水流提供一个均匀的管壁，导流结构的上游侧为流线型的，以防止涡流的形成。用螺栓把导流板固定在灯泡体上，导流板之间通过螺旋撑条相连以提高刚度。相对于进口竖井和主支柱而言，导流板是独立支撑的，可以进行轴向运动。

为了检查和维修发电机进口竖井与水轮机支柱之间的空间，导流板上还设有检修孔。

7.3.5 旋转部件

图 7.5 给出了旋转部件，包括转轮、水轮机轴、轴封凸轮、锁紧圈、耐磨环、甩油环、水轮机侧轴承甩油环、反馈机制和润滑油管系，以及从旋转部件到固定部件的受油器。

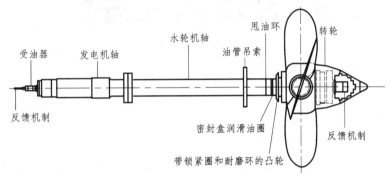

图 7.5 灯泡贯流式水轮机的旋转部件

7.3.5.1 转 轮

灯泡贯流式水轮机的转轮与轴流转桨式水轮机的转轮类似，通常有 3~5 个不锈钢叶片。叶片上配有法兰，用来连接耳轴和转臂。

带动叶片转动的接力器通常位于轮毂内，如图 7.6 所示。接力器由固定的活塞、轴向移动的接力器缸和支撑连杆组成，连杆和叶片转臂位于轮毂内。

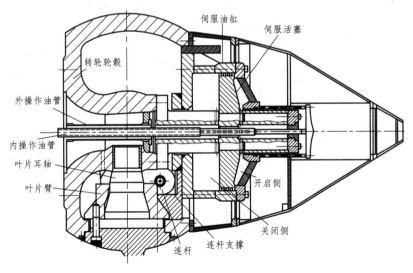

图 7.6 转轮轮毂

207

图 7.7 为灯泡贯流式水轮机转轮的一张照片。

图 7.7　车间内的转轮装配

7.3.5.2　水轮机轴

水轮机轴是由平炉钢锻造而成的，两端都配有法兰。水轮机轴的一端与转轮轮毂相连，另一端与发电机轴相连。这些接头都是纯摩擦接头。

7.3.6　轴封盒

目前使用的轴封盒有几种类型。图 7.8 给出了特别适用于灯泡贯流式水轮机的一种轴封盒。

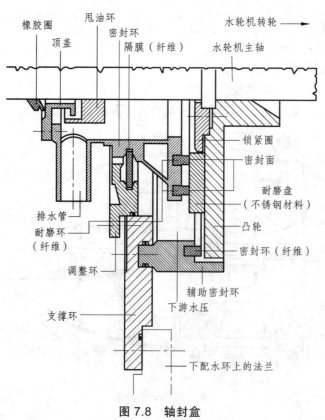

图 7.8　轴封盒

该轴封盒的径向密封面由不锈钢硬化耐磨盘和两个聚四氟乙烯纤维耐磨环组成。用螺栓把耐磨环固定在主轴上的凸轮上。耐磨环粘在密封环上。密封环是可以移动的，在调整环内由隔膜进行支撑。

隔膜使密封环可以轴向移动 5~6 毫米。当机组加载时，轴向移动对于主轴向下游侧移动是非常必要的。此外，必须为密封面磨损提供余量。

用螺栓把调整环固定在支撑环上，通过双动式顶起螺丝可以改变调整环的轴向位置。根据耐磨环的磨损范围，密封盒的调整范围在 8~10 毫米。

辅助密封位于支撑环内部。通过推拉式顶起螺丝可以使辅助密封靠近或远离凸轮。当辅助密封环与凸轮相接触时，不需要放空机组就可以拆卸耐磨密封环。

渗漏到密封盒内的水由排水管排到废水收集池中。

在主轴上安装了一个甩油环以防止水沿轴渗漏。在主轴的上游端安装了一个橡胶圈，以对密封盒盖进行密封。

密封盒中有 4 个弹簧，使耐磨密封环紧贴密封面，防止平衡系统不起作用时发生渗漏，如水轮机充水时。

7.3.7 水轮机轴承

图 7.9 给出了一个轴承设计的实例。轴承很坚固，操作也很简单。轴承的维护通常只是更换润滑油。

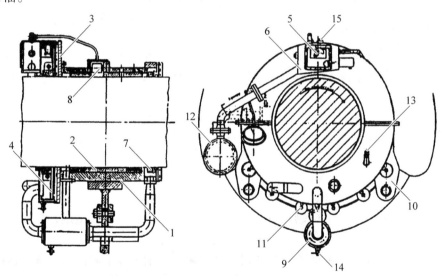

图 7.9 水轮机轴承

1—轴承座；2—轴承垫片；3—集油罩；4—旋转储油器；5—集油盘；6—浮动框；7—甩油环；
8—自动启动式润滑装置；9—沉积物收集器；10—U 形辊架；11—楔块；
12—油箱；13—注油孔；14—放空阀；15—检修注油塞

如图 7.2 所示，在内配水环内通过两个轭架和两个撑杆支撑轴承座（1），通常把轴承座安放在 6 个楔块上。通过这些楔块的轴向移动，可以垂向调整轴承座。轴承座可以水平分为两部分。

轴承垫片（2）由上下两部分组成，如图 7.9 所示。在轴承座内，轴承垫片并不是固定的，垫片上面安装了一个径向定位销钉以防止垫片转动。下部支撑垫片的表面与上部轴承垫片的两个端面都采用巴比特合金作为内衬。

用螺栓把集油罩（3）固定在轴承座的上游面。储油器安装在轴上。集油盘（5）和浮动框（6）都位于集油罩内。浮动框上设有一个窗口用来观察水轮机运行时润滑油的循环过程。

主轴上安装了一个甩油环（7）以防止润滑油沿着主轴从下游侧的轴承中渗出。

自动启动式润滑装置（8）安装在轴承座的上面。润滑装置中配有一个容器，该容器在水轮机运行时充满润滑油。当主轴停止转动时，由主轴固定的支撑装置使润滑油储存在该容器内。一旦主轴开始转动，支撑装置被移除，容器便开始倾斜。容器中的润滑油将被配送到轴承的表面上。

当主轴和储油器开始转动时，储油器从集油罩的下半部分中吸取润滑油。一旦油膜厚度足够大，集油盘将提取润滑油并把润滑油输送到浮动框中，然后送到油箱和轴承垫片中。旋转的主轴继续把润滑油配送到轴承的表面上。

通常情况下，油箱中循环的润滑油要比实际需要的润滑油多。因此需要提供一个旁通管把多余的润滑油送回集油罩的上面。该旁通流量由浮动框（6）内的一个浮动开关进行控制。

为了使润滑油的总量大于集油罩的容量，在轴承旁设置了一个油箱（12）。

沉积物收集器（9）位于轴承座下面。润滑油在轴承上循环的过程中截留的所有污垢颗粒在润滑油回流到集油罩之前都应在此进行分离。

轴承上设有各种注油孔、油位指示器、油位浮标和温度传感器。

7.3.8　反馈机制和操作油管

反馈机制和操作油管都位于轴的中心。操作油管由内外两个同心油管组成，油管贯穿整个主轴的长度。内操作油管与上游侧的受油器相连，由外套管进行支撑，通过轭架与转轮接力器缸相连。

内操作油管可以轴向移动，跟随接力器动作。在内操作油管的上游侧安装了指针，指针在测量尺上移动，指示接力器在任何时刻的机械位置。外操作油管通过法兰分别连到转轮轮毂、水轮机轴和发电机轴上。

7.3.9　受油器

受油器位于发电机轴的上游侧，由一个固定部件和一个转动部件组成，分别为配油套筒和配油耳轴。

配油套筒固定在发电机轴端的封装体内，并提供与供油、回油和漏油的管道连接。配油套筒上安装了带有测量尺的支架，可以读取转轮接力器的位置。

7.3.10　导叶操作机构

有两种不同的传动系统可以用来驱动导叶。Kværner Brug 设计了一种每个特定的叶片都有自己的接力器的一种传动系统，如图 7.10 所示。

通过连接环，可以实现各导叶接力器操纵阀的同时动作。操纵阀控制导叶的开启或关闭。

图 7.10 中的 A—A 剖面图解释了接力器供油和排空的整个过程。

高压软管与机组的油压系统相连。

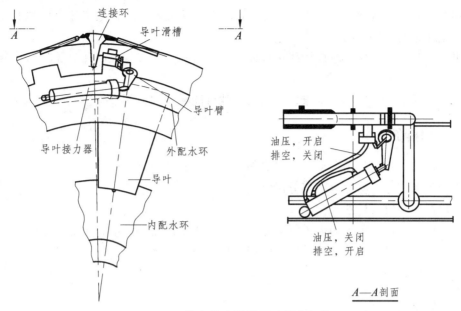

图 7.10 单个接力器的导叶操作机构

这个系统的优点在于即使一个导叶被卡住，其他导叶仍然可以动作且不会被损坏。同样适用于在关闭的过程中两个叶片被异物卡住的情况，其他叶片可以关闭而不会被损坏。如果需要，可以单独操作某些叶片，因此可以很容易的把卡在导叶系统中的异物冲走。

该系统的缺点在于组装复杂，需要大量的调整工作。然而该系统的总成本与另外一个导叶操作系统——控制环系统大概是相同的。

该控制环系统与混流式水轮机中由连杆和转臂组成的控制环系统是完全相同的。图 7.11 给出了一个带有 3 个接力器的控制环系统。由于导叶的锥形布置，必须为转臂和连杆系统配备可以进行大角度动作的球面轴承。

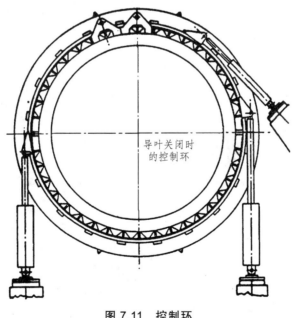

图 7.11 控制环

采用摩擦接头连接转臂与叶片。这样可以避免一个或几个叶片被卡住或异物卡开在叶片之间时，损坏某些部件。

如果相邻叶片被卡住，摩擦接头可以使叶片转臂随着连接到控制环上导叶的剩余部分一起动作。

该系统的缺点在于控制环很重，轴承的松弛可能导致调节不准确。

7.4 组装和拆卸

在灯泡贯流式水轮机的大型主要组成部分中，只有转轮需要进行拆卸。在轴向可以把转轮室水平分为两部分。移除上面部分后，则可以进入转轮。

导叶的拆除需要大量的工作。导叶及导向面的修理需要在电站内进行。

轴承和密封盒可以很容易地进行拆卸。通过采用检修密封，不必放空管道中的水就可以移除密封盒。然后可以在导叶和转轮室的周围搭建必要的楼梯。

导流板的位置要根据灯泡体和水管外壁进行调整。所以发电机灯泡体和压力水管安装完后，应尽快安装导流板。

最后安装发电机舱顶板和盖板。

8 水轮机调速

8.1 简 介

水轮机在驱动发电机或交流发电机时，最重要的一点为主轴的转速及水轮机转轮的转速必须保持恒定。否则将改变电力输出的频率。但当用电量变化导致电力负荷变化时，水轮机的转速应该自动改变。这是因为主轴上的外阻力矩改变了，而水流流过水轮机时，因动量变化产生的驱动力矩没有改变。例如，当负荷增加时，水轮机的转速降低，反之亦然。因此通过调整输入水轮机的能量可以相应地保证转速恒定。通常通过改变流过水轮机的流量来实现输入能量的改变——负荷增加时流量增加，负荷降低时流量降低。根据负荷调整流量的过程被称为水轮机的调速。

一般通过流量控制实现转速的调整。通过活动导叶的开关，根据负荷改变流量来控制混流式水轮机。为了保证负荷稳定时活动导叶的位置固定不变，调速器的执行元件要克服液压力和摩擦力。为此，大多数调速器都采用液压执行机构。轴流转桨式水轮机的特点是活动导叶和水轮机叶片都可以调整。通过两者调整的相互配合获得最大效率。对于每一个活动导叶的位置，都要选择一个水轮机叶片的位置使输出功率与水量之比最大化。另一方面，冲击式水轮机更容易控制。这是因为不需要改变压力水管中的流量就可以偏转射流或采用辅助射流绕开产生动力的射流。进而为流量调整到新的功率状态提供了更长的延迟时间。控制流量的喷杆或针阀可以缓慢的关闭，比如说 30 至 60 秒，从而减小了压力水管中的压力上升幅度。

8.2 水轮机调速的要求

8.2.1 频率和负荷调整

在独立电网中运行时，调速器应该能够保证发电机组的稳定性。通常，机组在部分负荷到满负荷的情况下，都应能够稳定运行。在这种运行模式下，调速器应把频率的偏移控制在一定范围内。

刚性系统中负荷的调整是最常见的操作模式。单个机组对电网系统频率的影响很小。调速器把负荷调整到期望值。作为频率变化的函数，负荷的变化与永态转差率的设置无关。

手动模式是一种特殊的操作模式，通过机械液压负荷限制器，手动的控制活动导叶的开度。这种模式只能用来控制负荷。

8.2.2 开机和停机程序控制

开机时，机组应能够快速平滑地达到额定转速。可以手动或自动开机。只有当所有的过载启动条件都满足后，才能引水。

关机时，应尽快地关闭进水，但是关闭速度还受到引水隧洞和调压井系统中压力上升幅度的限制。出于安全的考虑，停机信号会同时发送给调速器的不同部分，如关闭负荷限制器或关闭应急操作关断阀。如果普通电压供应出现故障，关断阀也会动作。可以手动或自动发出停机命令。

8.2.3 切断、甩负荷

切断是指打开发电机的主电路开关，从而把发电机从电网中分离出来，水轮机的功率输出将造成机组的转速上升。调速器的作用是关闭水轮机，但是关闭速度不能太快，要使产生的压力上升值在保证水平以下。

8.2.4 负荷限制

必须要能够根据外部条件对负荷进行限制。负荷限制设备可以手动或自动操作。

8.3 调速器的作用

与水力发电系统相连的大多数设备都对频率变化很敏感。因此，系统转速的控制是非常必要的。调整进入水轮机转轮的水量是调整和保持驱动发电机的转速恒定以及调整功率输出的一种常用手段。通过操作活动导叶或阀门来实现流量的调整。这需要一个控制活动导叶的机构，即调速器或调速器系统。水轮机的调速器是控制和调整水轮机功率输出，尽快消除水轮机功率与电网负荷之间偏差的一种设备。

水轮机调速器必须能够实现以下两个目标：

① 在任何电网负荷以及压力水管中现行的压力下，都能保证转速稳定、水轮发电机组恒定。
② 甩负荷或紧急停机时，必须结合机组转速上升的限制以及压力水管中压力上升的限制关断水轮机的进水。

电网负荷的变化将导致水轮机功率输出与负荷之间产生偏差。负荷降低时，多余的能量使转速增加，机组中水体转动速度加快。接下来，调速器减少进入水轮机的水量意味着减少压力水管中的水量，增加压力。

水轮机的调速要在考虑转速上升和压力上升限制的前提下，完成稳定性调节。为了把转速上升控制在甩负荷给定的极限范围内，进水阀门关闭的速度必须等于或大于一定值。为了控制压力水管中的压力上升，情况正好相反，如进水阀门的关闭速度必须等于或小于一定值才能尽量减小压力上升，使其在规定值以内。

如果电站中单独控制不能满足上面的两个要求，可以给调速器配备双重控制功能，一个控制转速上升，另外一个控制压力上升。高水头的切击式水轮机和混流式水轮机都采用这种调速器。

切击式水轮机的调速原理为：
——设置喷嘴中针阀关闭的速度以满足规定的压力上升。
——采用折向器临时偏转射流，保证转速上升值不超出可以接受的范围。

混流式水轮机的调速原理为：
——设置活动导叶开度的关闭速度，以满足转速上升的限制。
——通过旁通控制阀尽可能多地把流量转走，保证压力水管中的压力上升在给定值以下。

8.4 具体的水轮机调速设备

8.4.1 切击式水轮机的双重控制

通常采用双重控制对切击式水轮机进行调速，如喷嘴中的喷针和折向器。

负荷变化较小时，喷针调节控制本身就可以满足控制要求。然而快速甩负荷时，要采用折向器控制并限制转速的上升。接力器使折向器做旋转运动，使射流偏离转轮。

由程序控制的喷嘴根据折向器接力器的动作调整喷针的位置，直到流量与新的功率/负荷平衡适应为止。达到平衡后，折向器逐渐远离射流，回到射流边缘的停止位置。

通过调速器面板上的主控制阀控制折向器的动作。通过连接杆、转臂和连杆把接力器的动作传递给折向器。

接力器动作的反馈信号传递给控制阀。

每个喷针的程序控制都是通过凸轮盘实现的，凸轮盘由折向器的接力器驱动。凸轮特有的形状作为喷针控制阀的输入函数，可以打开控制阀把喷针调整到正确的位置上。

在阀门开度到达中间位置以及根据参考喷针移动到正确的位置之前，喷针动作的反馈信号要一直控制阀门的开度。

喷针的控制柜应尽可能靠近喷嘴以简化反馈系统，缩短油压管的长度。

在切击式水轮机调速器的最新设计中，都采用电子器件和单独的电液伺服系统来实现折向器和喷嘴的控制功能。在反馈系统中也常采用电子器件。这样大大简化了带有机械转臂和连杆的接力器的构造。

8.4.2 混流式水轮机的旁通控制

8.4.2.1 功能与总体布置

高水头混流式水轮机电站中甩负荷后，必须要把部分瞬时流量引离水轮机。这样可以迅速关断水轮机的引水，并延迟压力水管中主流的关断时间，把压力上升降到最小。

通常在蜗壳中增设一根支管来实现流量的转移，如示意图 8.1 所示。支管系统中安装了旁通阀，通过水轮机调速器控制旁通阀的进水。导叶的动作控制阀门的开度。导叶动作与流量旁通的联合控制使压力上升与转速上升在甩负荷时得到控制。

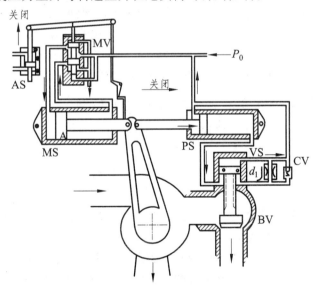

图 8.1 带有调节系统的旁通阀

AS—辅助接力器；MV—主阀；MS—主接力器；PS—中间接力器；VS—阀门接力器；
BV—旁通阀；CV—止回阀；d_1—速度控制孔；P_0—油压

通过阀门的水流进入消能器，然后流入水轮机尾水管，如图8.2所示。

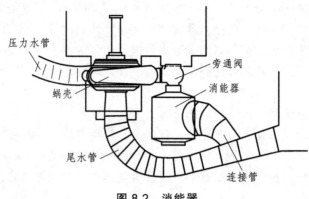

图8.2 消能器

挪威的一家水轮机制造商采用这种旁通阀系统已有很多年了，这种设备的长期使用已经证明该系统是简单可靠的。

该控制系统的设计要点为确保压力上升的完全控制，即便阀门操作失败也不例外。在这种情况下，导叶的关闭速度为压力上升允许范围内的给定速度。此时转速比正常值稍高一点，但并不太严重，因为发电机可以承受短时间内机组的飞逸转速。

8.4.2.2 阀门控制系统

图 8.1 为阀门系统和主接力器控制的示意图。辅助接力器 AS 通过主阀 MV 控制导叶接力器。双动式接力器 MS 通过控制环转动导叶。中间接力器 PS 与控制环相连，复制导叶接力器 MS 的动作。

通过孔口 d_1 的供油压力 P_0 给中间接力器 PS 加压。旁通阀接力器 VS 与中间接力器 PS 是通过液压连接的。在静止的条件下，由于活塞两侧的面积不同，旁通阀接力器处于关闭位置。

导叶接力器 MS 关闭时，来自于中间接力器 PS 的润滑油流过孔口 d_1 进入收集器。如果关闭速度超过一定值，孔口 d_1 使阀门接力器 VS 开放区域的压力增加，VS 开启。

为了避免限制导叶的动作，与孔口 d_1 并联连接了一个止回阀 CV。

阀门接力器 VS 相对于中间接力器 PS 的容积大小取决于满流量时旁通阀的开孔大小。

8.4.3 轴流转桨式和灯泡贯流式水轮机的双重控制

通过导叶叶栅控制与转轮叶片控制的最佳配合，可以使轴流转桨式水轮机和灯泡贯流式水轮机的效率达到最优。

可以采用机械液压或电动液压操作实现这两个控制功能的配合。联合装置通常位于水轮发电机组的顶部或旁边。

机械液压式联合装置集成在转轮叶片的控制系统中，由以下部分组成：主阀、反馈机制、联合控制功能曲线按钮、受油器的连接管路。

联合控制功能是指：通过主阀为转轮叶片的操作提供润滑油；根据控制函数曲线盘片确定转轮叶片的位置，盘片由导叶控制进行调节；反馈主阀的阀芯位置。

电动液压联合装置接收导叶叶栅位置的电反馈，这种情况下联合控制功能是以电子方式实现的。电动液压伺服阀操作接力器。

9 水轮机选型

每一个水电站都是独一无二的。把一个水电站与另一个水电站相比就像把橘子与苹果相比一样。为了最大化的利用河流的流量,水轮机的选型是非常重要的。如果水轮机选错了,则整个水电站的显著特征将被改变。

水电项目所在地的水头、流量和功率需求确定之后,可以采用图 9.1 来协助确定适合该项目的水轮机。横轴表示流量 Q,单位为立方米每秒。纵轴表示可用净扬程 H,单位为米。与流量和水头呈 45 度角的第三个坐标轴表示功率输出 P,单位为兆瓦。

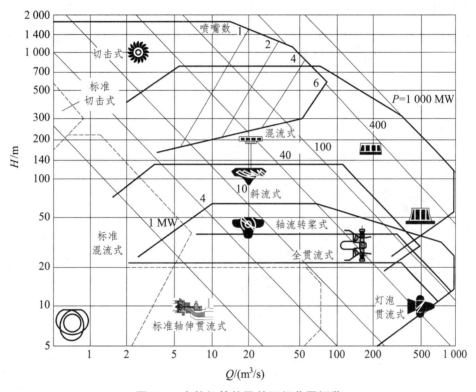

图 9.1 水轮机转轮及其运行范围概览

水轮机可以输出的功率为可用水头和流量的函数:

$$P = \eta \rho g Q H \tag{9.1}$$

式中,η 为水轮机的效率,ρ 为水的密度(千克每立方米),g 为重力加速度(米每平方秒),Q 为流量(立方米每秒),H 为静水头(米)。

水轮机的可用静水头决定了适用于一个特定厂址的水轮机的类型。流量决定了水轮机的容量。比转速通常用来对水轮机进行分类,并确定每类水轮机的特点。

水轮机的选型取决于以下几个因素:
① 厂址的特点;
② 水电项目的水头;
③ 项目的可用流量;

④ 发电机转轮的理想（预定）转速；
⑤ 水轮机在流量降低时运行的概率。

9.1 转 速

在指定水头的情况下要产生给定的功率，水轮机和发电机机组的转速越高（可行的）则期初投资越低。然而，转速还受到机械设计、空化趋势、振动、最高效率下降或整体效率损失（由于功率效率曲线的高效区很窄）的影响。转速太高还会减小水轮机可以安全运行的水头范围。

9.2 比转速

比转速通常是水轮机选型的基础。比转速是指给定水轮机同比缩小后，导叶全开时在一米水头下发出一公制马力时水轮机的转速，单位为转每分。低比转速与高水头相关，高比转速与低水头相关，如图9.2所示。此外，对于一个给定的水头，存在一个很宽的比转速范围。

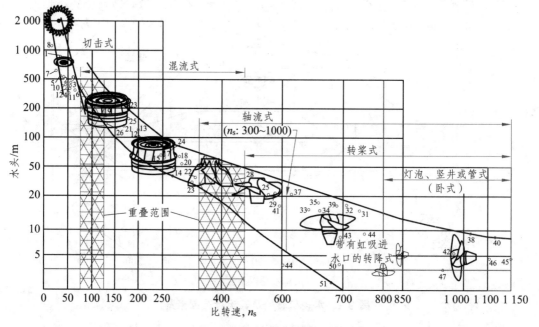

图9.2 比转速与水头

不同类型水轮机的比转速（公制马力单位）范围如下：
轴流定桨式水轮机：300～1000；
轴流转桨式水轮机：300～1000；
混流式水轮机：65～445；
切击式水轮机：每个喷嘴16～20，多喷嘴式水轮机的功率按比例增加；
双击式水轮机：12～80。

给定水头时，选择的比转速越高，水轮机和发电机的尺寸越小，投资费用也越小。然而，比转速越高，反击式水轮机需要安装的越低，相应产生的费用可能抵消之前节约的投资费用。选择一个经济的比转速还要考虑电能的价格、发电厂利用率、利率以及分析期限。常用的比转速的数学表达式是基于功率的（英制），如下：

$$n_s = \frac{n_r \sqrt{P_r}}{H_r^{5/4}} \tag{9.2}$$

式中　n_r——转速，转/分；
　　　P_r——导叶全开时的功率，公制马力；
　　　H_r——额定水头，米。

比转速的值定义了每种类型和尺寸的水轮机近似应用的水头范围。低水头的机组通常比转速较高，而高水头的机组通常比转速较低。比转速确定后，就可以采用图 9.2 来确定一个特定项目可以选用的水轮机类型。冲击式水轮机在相对较窄的一个比转速范围内才是高效的，而混流式水轮机和轴流式水轮机的可用范围很宽。

需要注意的是图 9.2 中不同类型设备的应用范围有重叠。这意味着在重叠范围内不止一种类型的机组可以高效的运行，此时最终的选择取决于其他因素，如发电机转速和空化。

9.3　水轮机效率

了解水轮机效率的变化趋势对水电站机组的选择很有用处。图 9.3 给出了不同类型水轮机的典型效率曲线。

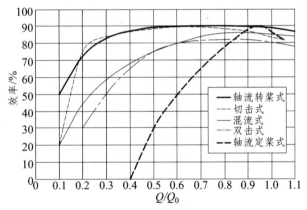

图 9.3　不同类型水轮机在流量降低时的效率曲线

这些曲线用来说明在设计水头下负荷变化时水轮机效率的变化。可以对比不同水轮机在设计工况下和流量降低工况下的相对效率。需要着重考虑的一个因素为是否需要水轮机在很宽的负荷范围内工作。

9.3.1　冲击式水轮机

冲击式水轮机通常在 1000 英尺以上的水头下工作时是最经济的，但是对于小型机组以及过载保护在其中很重要的情况下，冲击式水轮机也可以用于较低水头。由于喷嘴的设计形式，切击式水轮机在很宽的功率载荷范围内都可以高效的运行，而双击式水轮机在设计流量一半以下的流量运行时，效率急剧下降。

9.3.2　反击式水轮机

反击式水轮机的几何形状是固定的，随着负荷的变化，效率变化很明显。混流式水轮机可以运行的流量范围为最佳效率流量的 50%～115%。低于 40% 时，效率低下、不稳定、运行困难使长时间的运行极不明智。流量的上限受稳定性、发电机额定值和温度上升的影响。水头的范围近似为设计水头的 60% 至 125%。

9.3.3 轴流式水轮机

轴流式水轮机的工作水头为 5～200 英尺，但是通常低于 100 英尺。与混流式水轮机相同，轴流定桨式水轮机的几何形状是固定的。由于效率随负荷变化较大，定桨式水轮机的工作流量范围应在最高效率流量的 75% 至 100% 之间。

然而，轴流转桨式水轮机和斜流式水轮机可以在很宽的工作范围内保持高效的运行，在低于设计流量下运行时，效率依然很高。轴流转桨式水轮机可以在最高效率流量的 25% 至 125% 之间运行。正常运转的水头范围为设计水头的 20% 至 140%。

选择结构简单但效率曲线比较尖的机组还是选择复杂昂贵但效率曲线较平缓的机组取决于电站的预期运行效果及其他经济因素。

9.4 水轮机安装高程和开挖要求

在水轮机选型和安装时，水轮机相对于尾水位的高程是一个很重要的因素。空化造成的水轮机点状腐蚀直接受到安装高程的影响。为了控制空化，通常把水轮机安装的更深，这样可以增加水轮机转轮上的压力，减少气泡的形成。很明显，水轮机埋深越大，电站施工的土建成本也越大。较高的水轮机转速也会增加水轮机空化的倾向，但是较高的转速可以减小设备尺寸，降低机组成本。通常水轮机的埋深要考虑设备成本与土建成本之间的平衡关系。

为了确定水轮机的空化性能，我们必须考虑水的蒸汽压力、大气压力、转轮相对于尾水位的安装高程，以及导叶全开时水轮机的最大水头。这些因素可以用托马空化系数来表示，关系式如下：

$$\sigma = \frac{H_a - H_v - H_s}{H} \tag{9.3}$$

式中　σ ——托马空化系数；

H_a——大气压力，英尺水柱（参见表 9.1）；

H_v——蒸汽压力，英尺水柱（参见表 9.2）；

H_s——转轮相对于尾水位的高程，英尺，测量点为混流式水轮机的喉口或立轴桨式水轮机的叶片中心线或卧轴水轮机转轮的尖端（见图 9.4）；

H——导叶全开时水轮机的总水头。

表 9.1　不同海拔时的大气压力

海拔（英尺）	H_a（英尺水柱）	海拔（英尺）	H_a（英尺水柱）
0	33.96	5 000	28.25
500	33.35	5 500	27.73
1 000	32.75	6 000	27.21
1 500	32.16	6 500	26.70

续表

海拔（英尺）	H_a（英尺水柱）	海拔（英尺）	H_a（英尺水柱）
2 000	31.57	7 000	26.20
2 500	31.00	7 500	25.71
3 000	30.43	8 000	25.22
3 500	29.88	8 500	24.74
4 000	29.33	9 000	24.27
4 500	28.79	9 500	23.81
		10 000	23.35

表9.2　不同温度时水的蒸汽压力

温度（华氏度）	H_v（英尺）	温度（华氏度）	H_v（英尺）
40	0.28	60	0.59
50	0.41	70	0.84

由于H_a和H_v相对恒定，所以任意给定装置的空化系数σ仅受H_s或H的影响。空化发生时的σ值被称为临界值，表示为σ_c。如果一个装置计算的σ值低于σ_c，则水轮机会发生空化。对于一个特定的转轮设计，采用模型试验的方法确定σ_c的值，如果缺乏测试数据，可以采用下式计算最低σ值：

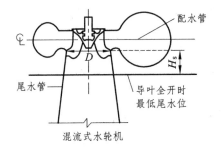

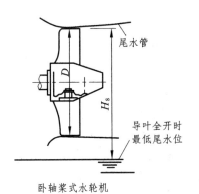

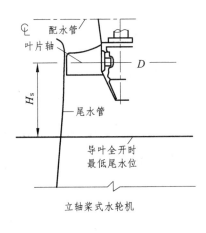

图9.4　水轮机安装高程系数的定义

D—水轮机转轮最小直径；H_s—导叶全开时，最低尾水位表面距立式机组最低转轮直径的距离或距卧式机组叶梢的距离

$$\sigma = \frac{n_s^{1.6}}{4325} \tag{9.4}$$

为了节约计算时间，把公式9.4绘成图，如图9.5所示。重新整理公式9.3，可以计算出转轮喉口的高程：

$$H_s = H_b - H_v - \sigma H \tag{9.5}$$

为了保证转轮的安装位置远远低于空化限制值，常规的做法是再减去一个额外的安全距离。根据经验，小型机组的安全值为3 ft。因此可以把公式9.5改写成以下形式：

$$H_s = H_b - H_v - \sigma H - 3 \tag{9.6}$$

由于不同制造商设置的安装高程不同，因此该方法只能用来初步确定水轮机的安装高程。这些计算只能用于项目的初始阶段，如厂房设计、计算所需开挖深度、选择不同的水轮机类型和设计。

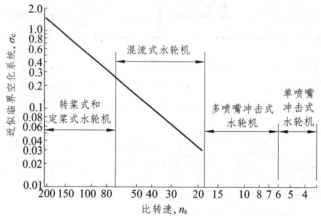

图9.5 临界系数

9.5 水轮机选型标准

9.5.1 水轮机选型的一般准则

① 20米以下的低水头，流量与负荷变化很大 —— 灯泡贯流式水轮机。
② 中低水头（80米以下），流量恒定 —— 轴流定桨式水轮机。
③ 中低水头（80米以下），流量与负荷变化很大 —— 轴流转桨式水轮机。
④ 30至550米的中高水头，水轮机流量变化在可以接受的范围内 —— 混流式水轮机。
⑤ 水头大于400米，流量与负荷变化的情况 —— 切击式水轮机。

下表中给出了根据厂址的条件，确定水轮机类型的因素。

表9.3 不同类型水轮机的适用范围特点

水轮机类型	大概的水头范围（米）	水头变化范围（%额定水头）	负荷变化范围（%额定负荷）	比转速范围（公制马力）	最高效率（%）
切击式	>300	120~80	50~100	15~65	92
混流式	30~400	125~65	50~100	60~400	96

续表

水轮机类型	大概的水头范围（米）	水头变化范围（%额定水头）	负荷变化范围（%额定负荷）	比转速范围（公制马力）	最高效率（%）
斜流式	50~150	125~65	50~100	200~400	95
轴流转桨式	10~60	125~65	40~100	300~800	95
轴流定桨式	10~60	110~90	90~100	300~800	93
灯泡贯流式	5~20	65~125	40~100	600~1200	95

9.5.2 重叠区水轮机的选择标准

（1）连续运行时的最低出力。

表9.4 不同水轮机的最低出力

水轮机类型	连续运行时的最低出力
切击式	30
混流式	50
轴流转桨式/灯泡贯流式	30
轴流定桨式	85
斜流式	40

（2）水轮机安装高程与开挖要求。
（3）运输要求（转轮）。

有时大型转轮的运输很困难，必须限制转轮的尺寸。最大外径为18英尺（5.5米）的转轮是铁路可以运输的最大转轮。更大的机组需要小心谨慎地分段施工和现场制作。现场制作的成本是很高的，只有当同时有几台机组可以分担设施的成本时才能采用。转轮可以分为两部分制作，在车间完全加工好后，在现场用螺栓连接起来。这也是很昂贵的，大部分用户都避免采用这种方法，原因在于转轮的完整性得不到保证。

（4）压力上升与转速上升要求。

表9.5 不同水轮机压力上升与转速上升的要求

水轮机类型	压力上升（%）	转速上升（%）
切击式	15~20	20~25
混流式	30~35	35~55
轴流转桨式/灯泡贯流式与轴流定桨式	30~50	30~65
斜流式	20~45	35~65

（5）水轮机效率。

① 水轮机在设计水头时性能是最好的。与混流式和轴流定桨式水轮机相比，切击式、轴流转桨式和灯泡贯流式水轮机的效率下降要小很多。因此，在水头重叠范围内选择水轮机时，

应该考虑厂址水头的变化范围。

② 水轮机的效率随负荷的变化而变化。混流式水轮机和轴流定桨式水轮机在部分负荷时效率降低比轴流转桨式水轮机和切击式水轮机要陡很多。因此水轮机长时间在部分负荷下运行的要求将影响在重叠范围内水轮机的选择。

因此，当水头范围同时适用于轴流转桨式水轮机和混流式水轮机时，水头和电力负荷的较大变化使轴流转桨式水轮机优于混流式水轮机，因为前者的整体效率高，所以发电量更高。

类似的，当重叠水头范围同时适用于混流式水轮机和切击式水轮机时，当负荷和水头变化较大时，切击式水轮机优于混流式水轮机。

③ 水轮机比转速高，则发电机转速高，可以降低发电机的成本。当考虑成本时，这一标准对水头重叠范围内的水轮机选型很重要。

（6）泥沙淤积的易感性。

9.6 机组台数的确定

通常情况下，对于给定的设施机组台数越少成本效益越好。但是，当流量变化很大时，为了充分利用水量，有必要采用多台机组。地址特征或现有的构筑物带来的空间限制等因素将影响机组的尺寸。

图 9.6 给出了多台机组并联时有效利用较低流量变化的原理。具有两台、三台或四台单机容量相同的机组的电站是最好的，这样的电站应能满足流量的任意变化。在设计分析阶段，如果可以采用标准化的机组时，可以把一台全容量机组，两台或多台大小相同的机组，两台或多台大小不同的机组作为备选方案，从中选定最优的设备类型。

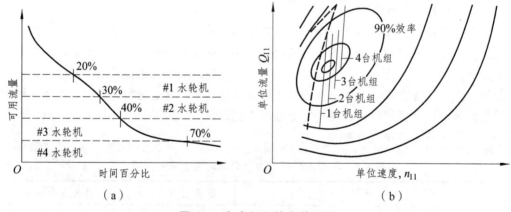

图 9.6 多台机组的有效利用

10 离心泵的分类及组成部分

离心泵是流体系统中最常见的一个组成部分。离心泵被广泛使用的原因之一是它可以在很大的流量和扬程范围内工作。

10.1 按出流方向分类

可以根据水流流出叶轮的方向对离心泵进行分类。泵壳和叶轮的设计决定了水流流过离心泵的方式。水流流过离心泵的方式有径向流、轴向流和斜向流三种。

10.1.1 径流泵

在径流泵中，液体从叶轮的中心进入泵壳，然后沿着叶轮叶片以垂直于泵轴的方向被导出。典型径流泵的叶轮以及水流在径流泵内的流向如图 10.1 所示。

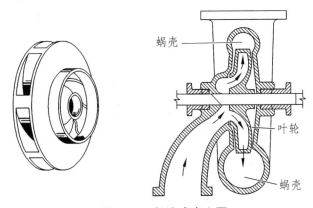

图 10.1 径流式离心泵

10.1.2 轴流泵

在轴流泵中，叶轮把液体沿着与泵轴平行的方向推出。有时候把轴流泵称为旋桨泵，因为轴流泵的运行原理与船只的螺旋桨基本上是一样的。典型轴流泵的叶轮及水流在轴流泵内的流向如图 10.2 所示。

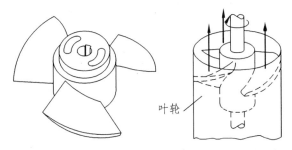

图 10.2 轴流式离心泵

10.1.3 混流泵

混流泵综合了径流泵和轴流泵的特性。当液体流过混流泵的叶轮时，叶轮的叶片推动液体远离泵轴，水流流出的方向与吸水方向的夹角大于 90 度。典型混流泵的叶轮及水流在混流泵内的流向如图 10.3 所示。

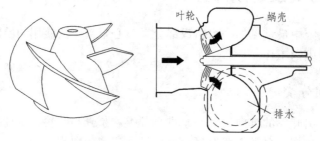

图 10.3　混流式离心泵

10.2　固定部件

10.2.1　泵壳和蜗壳

离心泵的基本组成部分包括固定的泵壳和安装在旋转轴上的叶轮。泵壳为离心泵提供一个压力边界，并正确的引导吸入流和排出流。泵壳上的进口和出口用来引导水流，泵壳上通常还要安装小的放水孔和排气孔，用来放空泵壳内的水进行维护或排出困在泵壳内的气体。

图 10.4 为一个典型离心泵的简图，图中可以看出吸水口、叶轮、蜗壳和出水口的相对位置。泵壳引导液体从吸水管进入叶轮的中心——吸入口。旋转叶轮的叶片对液体施加径向力和旋转力，把液体甩向泵壳的外周，并在一个叫蜗壳的地方进行收集。蜗壳是指围绕泵壳且横截面面积不断扩大的一个区域。蜗壳的作用是收集从叶轮周边甩出的高速液体，并通过扩大过水断面面积来逐渐降低流速。进而把动压头转换为静压头。然后液体从离心泵的压水管排水。

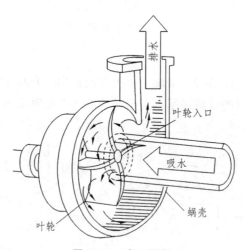

图 10.4　离心泵图

可以为离心泵设计两个蜗壳，每个蜗壳都在任意给定的时间内收集从叶轮周围 180 度甩出来的水。这种形式的泵叫双蜗壳泵（也可以称为剖分式蜗壳泵）。在一些应用中，双蜗壳可以把施加到泵轴和轴承上的径向力最小化，该径向力是叶轮周围的压力不平衡造成的。在图 10.5 对单蜗壳泵和双蜗壳泵进行了对比。

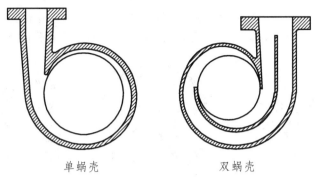

图 10.5　单蜗壳和双蜗壳

10.2.2　导　轮

部分离心泵泵壳内装有导轮，如图 10.6 所示。导轮是安装在叶轮周围的一些固定叶片。导轮的作用是通过设置一个更加逐渐扩张的区域为液体提供一个湍流较少的区域，进而降低流速，提高离心泵的效率。在设计导轮叶片时，应保证液体离开转轮后流过导轮时的过流面积不断扩大。过流面积的增加可以降低流速，并把动能转化为压能。

图 10.6　离心泵导轮

10.3　转动部件

10.3.1　叶　轮

可以根据进水口的数量以及叶轮叶片之间盖板的覆盖情况对离心泵的叶轮进行分类。

叶轮可分为单吸式和双吸式。在单吸式叶轮中，液体只能从一个方向进入叶片的中心。在双吸式叶轮中，液体可以同时从叶轮的两侧进入叶片的中心。图 10.7 所示为单吸式和双吸式叶轮的简图。

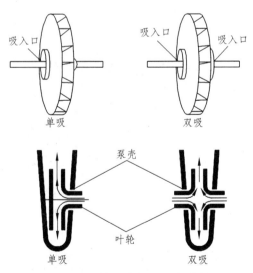

图 10.7　单吸式和双吸式叶轮

叶轮又可以分为敞开式、半开式和封闭式三种形式。敞开式叶轮仅包括安装在轮毂上的叶片。半开式叶轮仅在叶片的一侧装有圆形盖板（腹板）。封闭式叶轮在叶片两侧都有圆形盖板。封闭式叶轮还被称为闭式叶轮。图10.8为敞开式、半开式和封闭式叶轮的简图。

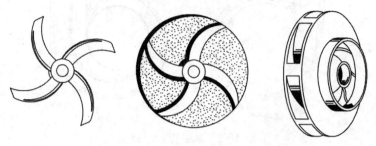

图10.8 敞开式、半开式和封闭式叶轮

10.3.2 泵轴

泵轴的作用是把原动机的输入功率传递给安装在泵轴上的叶轮。泵轴要承受几个应力：挠曲应力、剪切应力、扭曲应力、拉应力等。其中扭曲应力是最重要的，通常作为确定泵轴直径的基础。泵轴常用的材料是4140碳钢和不锈钢，如310，410或416。泵轴是水平放置还是垂直放置决定了是卧式泵还是立式泵。

10.4 轴封装置

离心泵中常用的轴封装置有填料盒、填料和水封环，如图10.9所示。

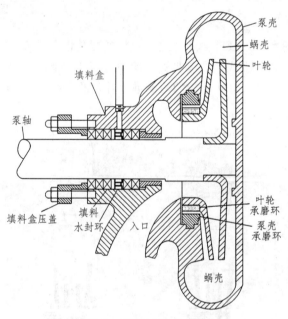

图10.9 离心泵的组成部分

10.4.1 填料盒

在几乎所有的离心泵中，驱动叶轮的旋转轴都要穿透泵壳的压力边界。因此在设计泵时，必须要控制泵轴穿过泵壳的位置上沿着泵轴的液体渗漏量。很多方法都可以用来对泵轴穿过泵壳的位置进行密封。需要考虑的因素包括被提升液体的压力和温度，泵的大小，被提升液体的化学和物理特性。

最简单的一种轴密封装置就是填料盒。填料盒是泵壳内围绕泵轴的一个圆柱形空间。空间内布满了一圈圈的填料。填料是放在填料盒内环形的或一股股的材料，用来形成密封以控制沿着泵轴的渗漏量。压盖用来压紧填料环。相应的，采用带有调节螺母的螺栓来固定压盖。如果拧紧调节螺母，螺母会向内挤压压盖，进而压缩填料。该轴向压缩使填料径向扩张，在旋转轴和填料盒内壁之间形成严密的密封。

高速旋转的轴与填料圈摩擦时会产生大量的热。如果不对填料进行润滑和冷却，填料的温度会升的很高，以至于破坏填料和泵轴，甚至可能破坏邻近的泵轴承。在设计填料盒时，经常允许少量可控的水可以沿着泵轴渗漏，对填料进行润滑和冷却。通过调整填料盖的松紧可以调节渗漏率。

10.4.2 水封环

并不是所有的离心泵的泵轴都可以采用标准的填料盒进行密封。有时泵的吸水侧可能为真空，这时水就不可能向外渗漏，或者提升的液体太热不能为填料提供足够的冷却。这时需要对标准的填料盒进行改进。

上述条件下对填料进行充分冷却的一个方法就是加装水封环，如图 10.9 所示。水封环是一个位于填料盒中心的中空穿孔环，用来从泵的出水侧或外部接收清洁的液体，并均匀的把液体分配到泵轴上进行润滑和冷却。进入水封环的液体可以冷却泵轴和填料，润滑填料，并密封泵轴和填料的连接处，当泵吸水侧的压力低于大气压时，还可以防止空气进入泵内。

10.4.3 机械密封

在某些情况下，不能用填料进行轴密封。另外一种常用的轴密封方法就是机械密封。机械密封包括两个基本部分：安装在泵轴上的旋转部分以及安装在泵壳上的固定部分。这些组件都有一个高度抛光的密封面。旋转组件和固定组件的抛光面之间相互接触，形成密封控制沿泵轴方向的渗漏。

10.5 承磨环

离心泵由旋转的叶轮和固定的泵壳组成。为了保证叶轮在泵壳内能够自由转动，在叶轮和泵壳之间设计了一个小的缝隙。为了尽量提高离心泵的效率，有必要尽量减少从泵的高压侧或出水侧通过该缝隙回流到低压侧或吸水侧的渗漏量。

在叶轮和泵壳几乎接触的地方会产生磨损或磨蚀。该磨损是由液体从这个狭窄的缝隙中渗漏而产生的磨蚀或其他原因造成的。磨损使缝隙变得越来越大，渗漏量也越来越大。最终，渗漏量大到不能接受的程度，需要对泵进行维护。

为了尽量降低泵的维护成本，大多数离心泵都安装了承磨环（减漏环），如图 10.9 所示。承磨环是可以更换的圆环，安装在叶轮和/或泵壳上，在叶轮和泵壳之间形成一个小的运行缝隙，避免磨损叶轮或泵壳的本体材料。在泵的使用期内，要定期的更换承磨环，从而避免更换价格较高的叶轮或泵壳。

10.6 轴向推力平衡装置

轴向推力在泵的稳定性方面具有很重要的作用。叶轮巨大的表面积使压力在叶轮盖板上积聚，在轴向对轴承产生很大的负荷。

不同形式叶轮的负荷特性也不同。没有盖板的叶轮不会产生轴向负荷，因为没有可以形成压差的盖板。半开式叶轮的轴向推力特性最差，因为只有一个盖板，盖板的一侧在整个表面上形成排水压力，而在另一侧随着水流沿着径向流出，压力从吸水压力增加到排水压力，如图 10.10 所示。这个压差通常可以达到成千上万磅，会直接施加到轴承上。闭式叶轮带有前后两个盖板，轴向推力平衡比较简单，但是仍然需要采用某种形式的轴向力平衡装置。

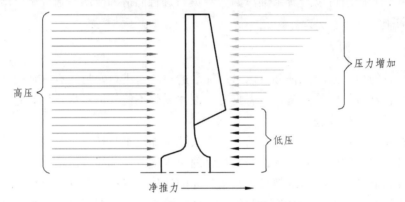

图 10.10　半开式叶轮上的轴向推力

在叶轮后盖板上安装的叶片就是一种轴向推力平衡装置。这些叶片把液体从叶轮的后面抽送到外径方向，因此降低了泵轴附近的压力，增加了出水直径方向的压力，这样就可以模拟叶轮前面的推力分布。这种设计方式通常应用于半开式叶轮。

另外一种比较流行的轴向推力平衡措施就是在叶轮的后盖板上开平衡孔，使叶轮后面的高压通过该平衡孔回流到吸水侧，如图 10.11 所示。大多数形式的推力平衡措施都会降低效率，增加泵的运行成本。但是这些平衡措施对水泵的机械完整性及对整个机组的稳定性是非常有用的。

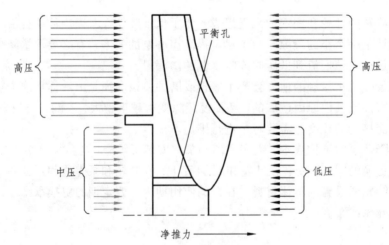

图 10.11　闭式叶轮上的轴向推力

10.7 多级离心泵

要使只有一个叶轮的离心泵在吸水口和出水口之间形成大于 150 磅每平方英寸的压差是很困难的，其设计费和建造费用都是非常高的。使一台离心泵形成高压的一个更加经济的方法就是在相同泵壳内的同一个泵轴上安装多个转轮。泵壳的内部流道把从一个叶轮甩出的水导向下一个叶轮的吸水口。图 10.12 给出了一台四级泵叶轮的布置情况。水从泵的左上侧进入，从左到右流过串联的四个叶轮。水沿着蜗壳流动，从一个叶轮的出水口流到另一个叶轮的吸水口。

把离心泵中一个叶轮以及其相关部件组成的整体称作泵的一级。大多数的离心泵都是单级泵，只有一个叶轮。有一个泵壳和七个叶轮的泵被称为七级泵或多级泵。

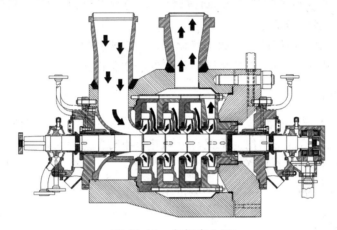

图 10.12 多级离心泵

11 离心泵的基本参数

为了了解离心泵的运行原理，必须首先了解离心泵的基本性能参数。

11.1 流量

11.1.1 体积流量

首先需要考虑的也是最重要的一点为离心泵是一种容积式机器。离心泵的流量是指单位时间内所输送的液体量。该流量实际上指的是体积流量。流量还可以用"capacity"和"discharge rate"来表示。流量的经典英制单位为加仑每分钟，公制单位为升每分钟或立方米每秒或立方米每小时。流量用字母 Q 表示。

值得一提的是，对于任何液体，如碳氢化合物、水或其他液体，在一个工况点上，任何泵的体积流量都是相同的，而质量流量随液体密度的变化而改变。

11.1.2 质量流量

质量流量 q 是指水泵出水管线中单位时间内排出的液体质量。质量流量随着流体密度的变化而变化。质量流量的常用单位为千克每秒和吨每小时。

质量流量 q 和体积流量 Q 的关系为：

$$q = \rho Q \quad (\rho —— 液体的密度)$$

11.2 扬程（水头）

扬程是水利工程中常用的压力单位。在分析静态或动态状态时，扬程是用来描述水泵设计压力与系统设计压力的参数。用字母 H 表示扬程。单位为英尺（公制单位为米）。

用下面的关系式表示：

$$扬程，英尺 = \frac{(压力[磅/平方英寸] \times 2.31)}{比重}$$

式中，比重为液体密度与水密度的比值。水的比重为 1.0。

静态系统中的压力为静水头（静扬程），动态系统中的压力为动水头（动扬程）。

11.2.1 系统水头（扬程）

流体流动系统中所有形式的能量都可以用英尺液柱来表示。这些不同水头的总和决定了系统的总水头或泵在系统中必须做的功。下面对不同的水头进行定义：

（1）总静水头（压头）是指吸水池液体表面与压水池液体表面之间的垂直距离。

（2）静排出压头（压水地形高度）是指泵吸入口中心线至压出液体表面的垂直距离。

（3）静吸入压头（负吸水地形高度），当吸水池水面高于水泵时采用。是指水泵吸水口中心线至吸水池液面之间的垂直距离。

（4）静吸上高度（正吸水地形高度），当吸水池水面低于水泵时采用。是指水泵吸水口中心线与吸水池液面之间的垂直距离。

上面四个术语在图 11.1 中进行了示意。

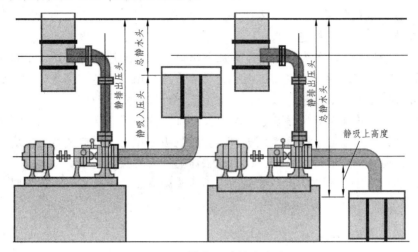

图 11.1 抽水系统的净水头

动水头由速度头、摩擦头和压头与上面的静水头一起来定义。

（5）速度头（h_v）是指液体以某个速度 v 流动时液体具有的能量。可以看成为了获得液体在管道中流动时相同的速度，液体需要下降的垂直距离。可以用以下关系式表示：

$$h_v = v^2/2g$$

式中，h_v 为速度头；v 为液体的流速，英尺每秒；g 为 32.2 英尺每平方秒。

速度头通常并不重要，在大多数水头较高的系统中都可以忽略。但是在水头较低的系统中是一个较大的系数，必须进行考虑。

（6）摩擦头（h_f）是克服液体在系统中的流动阻力所需的水头。摩擦头取决于管道的尺寸、状态和类型，管道配件的数量和类型，流量及液体的性质。

（7）压头，当抽送系统的吸水池或压水池内有压力且不是大气压时必须考虑。系统水头必须加上吸水池中的真空值或压水池中的正压值，反之，必须减去吸水池中的正压值或压水池中的真空值。如果存在真空，且真空值以英寸汞柱表示，可以用下面的公式计算相应的英尺液柱：

$$真空值，英尺 = \frac{英寸汞柱 \times 1.13}{比重}$$

在进行系统分析时，必须把该压力转换为英尺液柱，以确保所有单位的一致性。

上述各种形式的水头，即静水头、摩擦头、速度头和压头之和即为特定流量下系统的总水头。

11.2.2 水泵扬程

探讨水泵的运行状况时，经常采用一些术语用来描述动水头。换言之，水泵运行时是动态的。当流体在系统中流动时，泵送系统也是动态的，因此也必须按动态来进行分析。为了便于分析，采用下面四个动态术语。

（1）总动吸入压头（h_s）为静吸入压头加上进口法兰处的速度头，再减去吸水管线中的总摩擦头。通过测试获取泵的总动吸入压头，计算方法为进口法兰上压力计的读数，转换为英尺液柱，并按照泵的中心线进行校正，再加上压力计安装点的速度头。

（2）总动排出压头（h_d）为静排出压头加上出口法兰处的速度头，再加上排水系统中的总摩擦头。通过测试获取泵的总动排出压头，计算方法为出口法兰上压力计的读数，转换为英尺液柱，并按照泵的中心线进行校正，再加上压力计安装点的速度头。

（3）总动吸上高度（h_s）为静吸上高度减去进口法兰处的速度头，再加上吸水管线中的总摩擦头。为了计算总动吸上高度，把进口法兰处的吸入压力（进口法兰处压力计的读数）转换为水头并按照泵的中心线进行校正，再减去压力计安装点的速度头。

（4）系统的总动压头（TDH）是指总动排出压头减去总动吸入压头（如果吸水池水面位于水泵以上）。

$$TDH = h_d - h_s \text{（吸入压头）}$$

当吸水池水面位于水泵以下，总动压头为总动排出压头加上总动吸上高度。

$$TDH = h_d + h_s \text{（吸上高度）}$$

11.2.3 TDH 的预测

离心泵是把能量传递给液体的一种动力机器。液体流过叶轮时，通过改变流速进行能量的传递。液体流过泵壳或扩散器时，大部分的速度能都被转换为压能（总动压头）。

为了预测任意一台离心泵总动压头的近似值，必须经过两个步骤。

第一步，采用下面的公式计算叶轮出口直径处的流速：

$$v = \frac{RPM \times D}{229}$$

式中，v 为叶轮边缘的流速，英尺每秒；D 为叶轮出口直径，英寸；RPM 为转每分钟（叶轮的转速）；229 为常数。

第二步，叶轮出口直径或边缘处的速度能约等于泵的总动压头，把上式中的 v 替换到下式中，

$$H = v^2/2g$$

式中，H 为总动压头，英尺；v 为叶轮出口直径处的流速（英尺/秒）；g 为 32.2（英尺/平方秒）。

给定转速和叶轮直径后，离心泵可以把任何比重或重量的液体提升到给定高度。因此，在分析离心泵及其系统时，经常采用的术语为英尺水柱而不是压力。

11.3 功 率

在物理学中，功率的定义为单位时间内所做的功。在工程领域，功率的定义为做功的能力。功率的单位为马力和千瓦。

讨论功率时，需注意离心泵系统有三种不同的功率。分别为：水力功率、制动功率（轴功率）和驱动功率或电机功率。

11.3.1 水力功率

泵输出功率或水力功率或水功率（WHP）是指泵输送给液体的液体功率。用下式进行定义：

$$WHP = \frac{Q \times TDH \times 比重}{3960} \quad （马力） \quad 或 \quad WHP = \frac{Q \times TDH \times \rho}{367} \quad （千瓦）$$

式中，Q 为流量，加仑每分；TDH 为总动压头，英尺液柱；常数 3960 等于一马力的磅数或英尺磅（33000）除以一加仑水的质量（8.33 磅）。

11.3.2 制动功率

为了给液体提供一定的功率，必须为泵轴提供更多的功率以克服固有的损耗。泵的输入功率或制动功率（BHP）是指提供给泵轴的实际功率。该值大于水力功率，多出的值为泵的能量损失。由下式进行定义：

$$BHP = \frac{Q \times TDH \times 比重}{3960 \times 泵的效率}$$

制动功率通常由水泵的制造商提供，并在性能曲线上进行表示。因此，本书中所指的马力或功率均为制动功率。

11.3.3 驱动功率

原动机，也被成为驱动器，是把自然能源转换为功的机器。驱动功率是原动机的名义功率或铭牌上的额定功率。两种主要的离心泵驱动器为电动机和汽轮机。

11.4 效　率

离心泵的总效率为输送给液体的功率与泵轴的输入功率之比，用符号 η 表示。

$$\eta = \frac{水力功率}{制动功率} = \frac{WHP}{BHP} = \frac{Q \times TDH \times 比重}{3960 \times BHP}$$

由于在转换过程中存在部分动能损失，所以水泵并不能把全部的动能转换为压能。能量的损失或没有转换为有用功的能量主要体现在三个方面。水泵的效率是考虑了所有这些损失的一个系数，这些损失将在"泵的能量损失"一章中进行探讨。

11.4.1 水力效率

水力效率用来度量圆盘摩擦与冲击损失造成的功率消耗，圆盘摩擦与冲击损失的计算将在"离心泵及其系统的特性曲线"一章中介绍。圆盘摩擦（摩阻损失）是主要的功率消耗源，是指液体与叶轮盖板之间的摩擦。圆盘摩擦是转速与叶轮形状的函数。冲击损失是指沿着叶轮和蜗壳流动方向的快速改变造成的损失。

$$\eta_h = \frac{TDH}{H_{theo}}$$

式中，η_h 为水力效率；H_{theo} 为没有考虑任何损失时，离心泵的理论扬程。

11.4.2 容积效率

容积效率是用来度量承磨环、级间衬套、半开式叶轮的平衡孔和叶片间隙处产生的回流损失。容积效率等于水泵中实际流出的流量与没有渗漏时流量的比值。

$$\eta_v = \frac{Q}{Q_{theo}}$$

式中，η_v 为容积效率；Q_{theo} 为没有考虑任何渗漏时，离心泵的理论流量；Q 为离心泵的实际流量。

11.4.3 机械效率

机械效率用来度量密封或压盖填料和轴承处的机械摩擦损失。

$$\eta_m = \frac{N_{theo}}{BHP}$$

式中，η_m 为机械效率；N_{theo} 为叶轮传递给液体的全部理论功率。

$$N_{theo} = \frac{Q_{theo} \times H_{theo} \times specific\ gravity}{3960}$$

泵的总效率为以上三个效率的乘积：

$$\begin{aligned}\eta &= \eta_h \times \eta_v \times \eta_m \\ &= \frac{TDH}{H_{theo}} \times \frac{Q}{Q_{theo}} \times \frac{Q_{theo} \times H_{theo} \times 比重}{3960 \times BHP} \\ &= \frac{Q \times TDH \times 比重}{3960 \times BHP}\end{aligned}$$

尽管机械损失和容积损失很重要，但是水力效率是占比例最大的因素。

11.5 转速

旋转速度通常被简称为转速，转速的单位为转每分，通常用符号 n 表示。

11.6 气蚀余量（NPSH）

水泵的气蚀余量（NPSH）是指水泵吸入口的液体压力与液体蒸汽压力的差值，单位为液柱高度。在"离心泵的空化与气蚀余量"一章中将详细的进行探讨。

12 离心泵基本理论

本章介绍了离心泵进行能量转换的理论基础。离心泵工作时,能量以机械能的形式传递给泵轴。在叶轮内部,该能量被转换为内能(静压力)和动能(速度)。该过程可以用欧拉方程来解释。叶轮进出口的速度三角形可以用来解释泵的方程,计算无损失理论扬程和功率消耗。

速度三角形还可以用来预测转速、叶轮直径和宽度改变时泵的性能。

12.1 速度三角形

图 12.1 为进口和出口速度三角形的实例。图中 U 代表叶轮的切线速度,绝对速度 C 是相对于周边环境的流体速度。相对速度 W 是流体相对于旋转叶轮的速度。角 α 和 β 为流体的绝对速度和相对速度与切线方向的夹角。可以采用矢量相加的方法,使这些速度矢量形成叶轮进口和出口的速度三角形。

$$\vec{C} = \vec{U} + \vec{W}$$

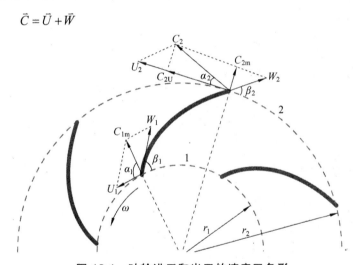

图 12.1 叶轮进口和出口的速度三角形

画出叶轮进口和出口的速度三角形后,采用欧拉公式可以计算出泵的特性曲线。

12.1.1 进 口

通常假设叶轮进水没有漩涡分离,即 $\alpha_1 = 90°$,如图 12.1 中位置 1 上的速度三角形。通过进口的流量和环形区域的面积可以计算出 C_{1m} 的值。

不同的叶轮形式(径向叶轮或半轴向叶轮)计算环形区域面积的方法也不同,如图 12.2 所示。对于径向叶轮:

$$A_1 = 2\pi \cdot r_1 \cdot b_1 \quad (\text{平方米}) \tag{12.1}$$

式中 r_1——叶轮进口边的半径(米);
b_1——叶槽进口宽度(米)。

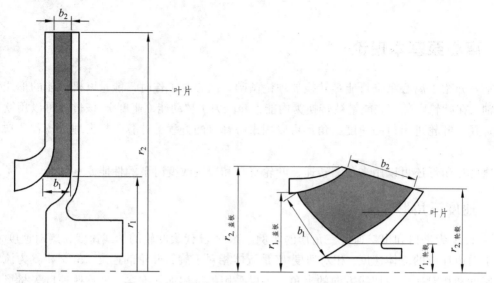

图 12.2 上面为径向叶轮，下面为半轴向叶轮

对于半轴向叶轮，

$$A_1 = 2\pi \cdot \left(\frac{r_{1,\text{轮毂}} + r_{1,\text{盖板}}}{2} \right) \cdot b_1 \quad （平方米） \tag{12.2}$$

所有水流都必须流过这个环形区域。C_{1m} 的计算公式为

$$C_{1m} = \frac{Q_{\text{叶轮}}}{A_1} \quad （米每秒） \tag{12.3}$$

切线速度 U_1 为半径和角速度的乘积：

$$U_1 = 2\pi \cdot r_1 \cdot \frac{n}{60} = r_1 \cdot \omega \quad （米每秒） \tag{12.4}$$

式中 ω ——角速度（度每秒）；

n ——转速（转每分）。

画出图 12.3 所示的速度三角形后，根据 α_1、C_{1m} 和 U_1，可以计算出相对进水角 β_1。如果进口没有漩涡（$C_1 = C_{1m}$），则

$$\tan \beta_1 = \frac{C_{1m}}{U_1} \tag{12.5}$$

12.1.2 出口

和进口类似，出口速度三角形在图 12.1 点 2 的位置上。对于径向叶轮，出口区域的面积为：

$$A_2 = 2\pi \cdot r_2 \cdot b_2 \quad （平方米） \tag{12.6}$$

对于半轴向叶轮，

$$A_2 = 2\pi \cdot \left(\frac{r_{2,\text{轮毂}} + r_{2,\text{盖板}}}{2} \right) \cdot b_2 \quad （平方米） \tag{12.7}$$

采用与进口同样的方法计算 C_{2m}，

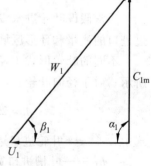

图 12.3 进口速度三角形

$$C_{2m} = \frac{Q_{叶轮}}{A_2} \quad （米每秒） \tag{12.8}$$

根据下式计算切线速度 U_2：

$$U_2 = 2\pi \cdot r_2 \cdot \frac{n}{60} = r_2 \cdot \omega \quad （米每秒） \tag{12.9}$$

在设计开始阶段，假设 β_2 与叶片安装度相同。采用下式计算相对速度

$$W_2 = \frac{C_{2m}}{\sin \beta_2} \quad （米每秒） \tag{12.10}$$

和 C_{2U}，

$$C_{2U} = U_2 - \frac{C_{2m}}{\tan \beta_2} \quad （米每秒） \tag{12.11}$$

据此可以确定并画出出口速度三角形，见图 12.4。

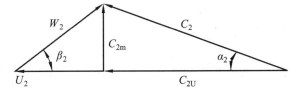

图 12.4 出口速度三角形

12.2 泵的欧拉方程

欧拉方程是泵设计时最重要的方程。可以采用不同的方式推导该方程。这里采用的方法为限制叶轮的控制体积以及描述进口和出口处的流体力和速度三角形的动量矩方程。

控制体积是用来建立平衡方程的一个假想的有限体积（叶槽）。可以为扭矩、能量及其他相关的流动参数建立平衡方程。动量矩方程是把质量流量和流速与叶轮直径相互关联起来的一个平衡方程。经常采用控制体积来描述叶轮，如图 12.5 中的 1 和 2 之间的控制体积。

我们所感兴趣的平衡是扭矩平衡。驱动轴传来的扭矩（T）相当于流体流过叶轮时产生的扭矩，此时质量流量为 $m = rQ$，即

$$T = m \cdot (r_2 \cdot C_{2U} - r_1 \cdot C_{1U}) \quad （牛米） \tag{12.12}$$

把扭矩与角速度相乘就可以得到轴功率 P_2。同时，半径与角速度的乘积为切线速度，$r_2 \omega = U_2$。所以

$$\begin{aligned} P_2 &= T \cdot \omega \quad （瓦） \\ &= m \cdot \omega \cdot (r_2 \cdot C_{2U} - r_1 \cdot C_{1U}) \end{aligned} \tag{12.13}$$

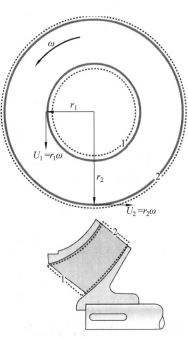

图 12.5 叶轮的控制体积（叶槽）

$$= m \cdot (\omega \cdot r_2 \cdot C_{2U} - \omega \cdot r_1 \cdot C_{1U})$$
$$= m \cdot (U_2 \cdot C_{2U} - U_1 \cdot C_{1U}) = Q \cdot \rho \cdot (U_2 \cdot C_{2U} - U_1 \cdot C_{1U})$$

根据能量方程，给液体增加的水功率可以写成叶轮内压力的增值 ΔP_{tot} 与流量 Q 的乘积，

$$P_{hyd} = \Delta P_{tot} \cdot Q \quad （瓦） \tag{12.14}$$

扬程的定义为

$$H = \frac{\Delta P_{tot}}{\rho \cdot g} \quad （米） \tag{12.15}$$

因此水功率的表达式可以写成

$$P_{hyd} = Q \cdot H \cdot \rho \cdot g = m \cdot H \cdot g \quad （瓦） \tag{12.16}$$

如果假设流动没有损失，则水功率与机械功率相等，

$$P_{hyd} = P_2$$
$$m \cdot H \cdot g = m \cdot (U_2 \cdot C_{2U} - U_1 \cdot C_{1U})$$
$$H = \frac{(U_2 \cdot C_{2U} - U_1 \cdot C_{1U})}{g} \tag{12.17}$$

该方程就是欧拉方程，表示的是叶轮的进口和出口的切线速度和绝对速度与扬程的关系。

如果对速度三角形采用余弦关系，泵的欧拉方程可以写成以下三个分量的总和：

- $\dfrac{U_2^2 - U_1^2}{2g}$ ——离心力引起的静扬程

- $\dfrac{W_2^2 - W_1^2}{2g}$ ——叶轮内流速变化引起的静扬程

- $\dfrac{C_2^2 - C_1^2}{2g}$ ——动扬程

$$H = \frac{U_2^2 - U_1^2}{2g} + \frac{W_2^2 - W_1^2}{2g} + \frac{C_2^2 - C_1^2}{2g} \quad （米） \tag{12.18}$$

如果叶轮内没有水流流过并且假设进水没有漩涡分离，则根据公式（12.17）扬程仅取决于切线速度，式中 $C_{2U} = U_2$：

$$H_0 = \frac{U_2^2}{g} \quad （米） \tag{12.19}$$

设计泵时，通常假设进水没有漩涡分离，即 C_{1U} 为零。

$$H = \frac{U_2 \cdot C_{2U}}{g} \quad （米） \tag{12.20}$$

12.3　叶片形状与泵的扬程曲线

如果假设进水没有漩涡分离（$C_{1U} = 0$），泵的欧拉方程（12.17）与公式（12.6）、（12.8）和（12.11）表明扬程随流量线性变化，且坡度取决于出水角 β_2：

$$H = \frac{U_2^2}{g} - \frac{U_2}{\pi \cdot D_2 \cdot b_2 \cdot g \cdot \tan(\beta_2)} \cdot Q \quad (\text{米}) \tag{12.21}$$

图 12.6 和图 12.7 给出了泵的理想扬程曲线与出水角 β_2 的关系。

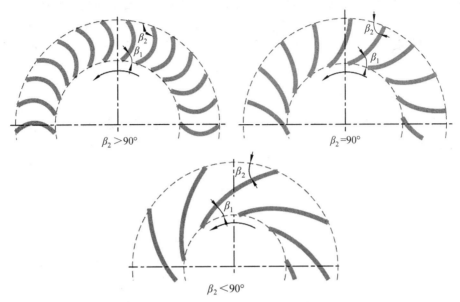

图 12.6 叶片形状取决于出水角 b2>90° 时的 H

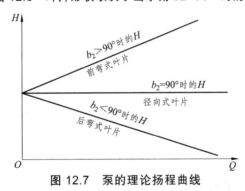

图 12.7 泵的理论扬程曲线

由于不同的损失、反旋、内旋等，实际的扬程曲线是弯曲的。

12.4 反 旋

在推导泵的欧拉方程时，假定液流随着叶片流动。但是由于进水角小于叶片安放角，实际上并非如此。这种状态被称为反旋。然而，进水角与安放角密切相关。对于叶片无限多、叶片无限薄的叶轮，流线与叶片的形状是相同的，如图 12.8 所示。

在叶片有限多、有限厚的实际叶轮中，液流不会完全沿着叶片的方向流动。叶轮出口的切向速度以及扬程都会因此而减小。下面给出了形成反旋的

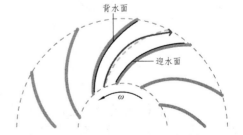

图 12.8 理想流线：虚线，实际流线：实线

一个可能成因。

液流在叶槽内流动时，叶轮叶片的迎水面和背水面上的压力和流速会产生差别。叶片迎水面上的压力偏高、流速偏低，而背水面上的压力偏低、流速偏高。结果是形成围绕叶片的环流，以及任意半径处速度的不均匀分布。这种情况下，出口处的平均流向从出口处的叶片安放角 β_2 变为另外一个角 β_2'，如图 12.9 所示。因此，出口处的切向速度分量 C_{2U} 降为 C_{2U}'，如图 12.9 中的速度三角形所示。切向速度分量的差值 ΔC_{2U} 被定义为反旋。反旋系数 σ_S 的定义为

$$\sigma_S = C_{2U}' / C_{2U} \tag{12.22}$$

考虑反旋系数 σ_S 后，泵传递给液体的工作扬程（欧拉扬程）变为 $\sigma_S C_{2U} U_2 / g$。反旋系数的典型值为 0.9。

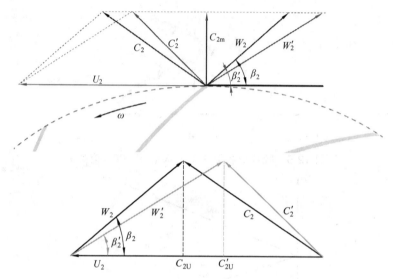

图 12.9　出口速度三角形，'表示反旋影响的速度

12.5　离心泵的比转速

离心泵比转速的概念与水轮机相同。区别在于离心泵中的相关参数为 n_s、H 和 Q，而水轮机中为 n、H 和 P。

对于泵来说

$$n_s = nQ^{1/2} / H^{3/4} \tag{12.23}$$

式中　n_s ——比转速（无量纲）；

　　　n ——泵的转速（弧/秒）；

　　　Q ——最高效率点的流量（立方米每秒）；

　　　H ——最高效率点的扬程（米）。

转轮的形状对比转速的影响也和水轮机类似。即径流式（离心）叶轮的比转速比轴流式叶轮的比转速低。叶轮而不是整个泵，特别是蜗壳的形状对比转速的影响很大。一般情况下，离心泵更适应中等流量、高扬程的情况，而轴流泵更适应大流量、低扬程的情况。与水轮机

类似，需求一定时，比转速越高，机组越紧凑。对于多级泵，比转速是指单级的比转速。

根据比转速可以推测出叶轮的形状以及泵扬程曲线的形状，如图12.10所示。

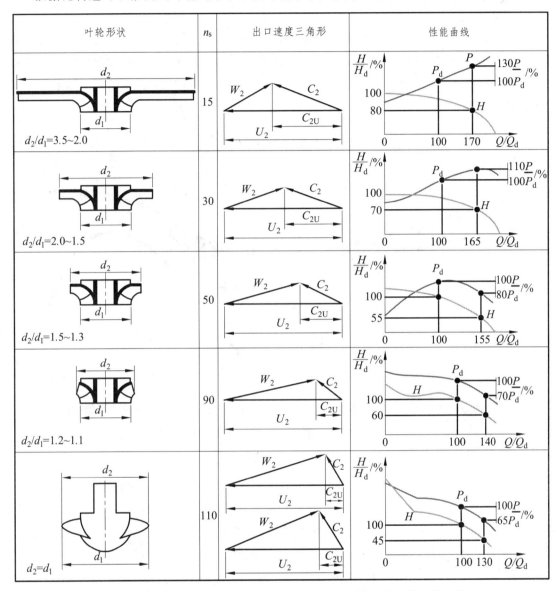

图12.10 叶轮形状、出口速度三角形和性能曲线都是比转速的函数

比转速低的泵被称为低比转速泵，这种泵的出口是径向的，而且出口直径比进口直径大。扬程曲线相对平缓，在整个流量范围内功率曲线的坡度都是正值。

相反，比转速高的泵被称为高比转速泵，这种泵的出口越来越趋向于轴向，与宽度相比，出口直径较小。扬程曲线通常是下降的，倾向于形成鞍点。随着流量的增加性能曲线下降。不同尺寸和不同形式的泵有不同的最大效率。

13 泵的能量损失

泵的欧拉方程对叶轮的性能进行了简单描述，并且忽略了能量损失。现实中，叶轮和泵壳内存在很多机械损失和水力损失，所以泵的性能要低于泵的欧拉方程所预测的值。这些损失使实际扬程低于理论扬程，并产生较大的功率消耗，如图 13.1 和 13.2 所示。结果导致效率的降低。

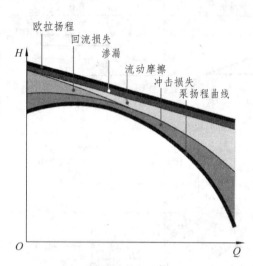

图 13.1 损失造成的理论欧拉扬程的降低

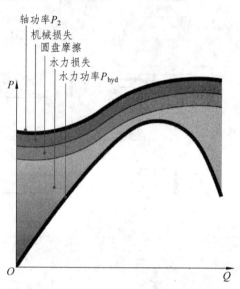

图 13.2 损失造成功率消耗的增加

13.1 能量损失的类型

泵的能量损失可以分为两种主要类型：机械损失和水力损失，这两种损失还可以进一步细分。机械损失可以进一步分为轴承损失和轴封损失，他们都会使功率消耗增加。水力损失可以进一步分为流动摩擦损失（摩阻损失）、混合损失、回流损失、冲击损失、圆盘摩擦损失和渗漏损失（容积损失）。前四种水力损失会降低水头，圆盘摩擦损失会增加功率消耗，而渗漏损失会减小流量。

泵的性能曲线中每种类型的能量损失都可以通过理论或经验计算模型进行预测。实际性能曲线取决于模型的详细程度，以及对实际泵型描述的程度。

从图 13.3 中可以看到泵内产生机械损失和水力损失的部件。这些部件包括轴承、轴封、前后腔密封、进口、叶轮和蜗壳或回流流道。本章中其他部分都将采用此图来说明每种类型的能量损失。

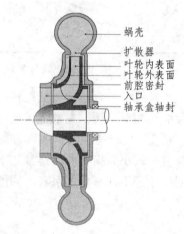

图 13.3 产生能量损失的部件

13.2 机械损失

泵的联轴器或驱动包括轴承、轴封和齿轮,不同种类的泵可能不同。这些部件都会产生机械摩擦损失。下面仅讨论轴承和轴封中的损失。

轴承和轴封损失也被称为附加损失,是由摩擦造成的。通常把该损失模拟为一个常数附加在功率消耗中。但是,损失的大小随压力和转速而改变。

下面的模型可以用来估计轴承和轴封的损失造成的功率需求的增量:

$$P_{机械损失} = P_{轴承损失} + P_{轴封损失} = 常数 \tag{13.1}$$

式中 $P_{机械损失}$ ——机械损失造成的功率需求的增量;
$P_{轴承损失}$ ——轴承的功率损失(瓦);
$P_{轴封损失}$ ——轴封的功率损失(瓦)。

13.3 水力损失

沿着泵内的流道方向水力损失在不断地增加。摩擦将造成水力损失,流体在流道内改变方向和流速时也将造成水力损失。截面面积的改变以及水流流过旋转的叶轮将导致流速和方向的改变。下面按水力损失的生成方式,对单个水力损失分别进行描述。

13.3.1 流动摩擦

流体与旋转叶轮表面和泵壳的内表面相接触的地方会产生流动摩擦。流动摩擦将产生一个压力损失,从而使扬程降低。摩擦损失的大小取决于表面的粗糙度和液体相对于表面的流速。

13.3.2 截面扩张处的混合损失

对于理想流体,压能、动能和势能之和为常数(伯努利方程),因此在泵内的截面扩张处动能被转换为静压能。转换时将伴随产生混合损失。

原因是横截面面积扩大时会产生速度差异,如图13.4所示。图中为一个截面面积突然扩大的扩散器,由于所有的水粒子不再以同样的速度移动,流体内的分子之间将产生摩擦,造成排出扬程损失。

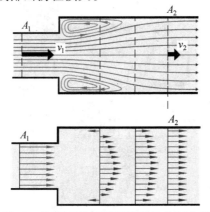

图 13.4 突然扩张后截面扩张处的混合损失

即便截面扩张后，速度剖面将逐渐趋于平稳后，如图 13.4 所示，还是会有一部分动能转化为热能，而不是静压能。

在泵的很多部位都存在混合损失：在叶轮出口流体流入蜗壳的位置或回流流道，以及导叶处。在设计液压元件时，提供一个小而光滑的截面扩张是很重要的。

13.3.3 截面收缩处的混合损失

水流接近几何边缘时形成涡流，将造成截面收缩处的扬程损失，如图 13.5 所示。

水流流过收缩断面时会分离。由于局部压力梯度的存在，水流不再平行于内表面，而是沿着弯曲的流线流动。这意味着水流的有效过流面积减小了，即收缩了。图 13.5 中标出了收缩面积 A_0。收缩使流速加快，因此水流流过收缩断面后，又必须减速以充满整个截面。此过程中将产生混合损失。在管道入口和叶轮入口经常产生截面面积收缩造成的扬程损失。可以通过把入口边修圆，来抑制水流分离，从而大大减小该损失。如果入口修的很圆，则该损失将很小。因此，截面面积收缩造成的损失通常是次要的。

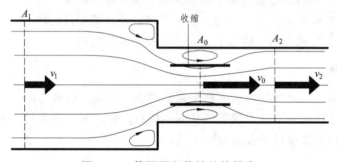

图 13.5 截面面积收缩处的损失

13.3.4 回流损失

通常当流量低于设计流量，即部分负荷时，在液压元件中将形成回流区。图 13.6 中给出了叶轮中回流的例子。回流区将减小水流流过时的有效过流截面面积。在具有较高流速的主流和流速接近于零的漩涡之间产生很大的速度梯度。结果是产生相当大的混合损失。

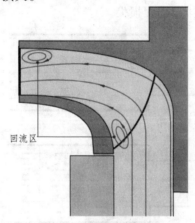

图 13.6 叶轮内回流的例子

在入口、叶轮、回流流道或蜗壳内可能形成回流区。回流区的大小取决于几何形状和运行工况。在设计液压元件时，在主要运行工况点处尽量减少回流区是很重要的。

13.3.5 冲击损失

当叶轮或导叶前缘处的进水角与叶片安放角不同时，将产生冲击损失。通常在部分负荷或存在预旋转时出现这种情况。

当进水角与叶片安放角不同时，在叶片的一侧会形成回流区，如图13.7所示。回流区在叶片前缘后造成水流收缩。收缩过后，水流必须再次减速以充满整个叶道，产生混合损失。

在偏离设计流量的工况下，在蜗壳舌部也会产生冲击损失。因此设计人员必须保证进水角和叶片安放角相互匹配，以尽量减小冲击损失。叶片边缘和蜗壳舌部修圆可以减少冲击损失。

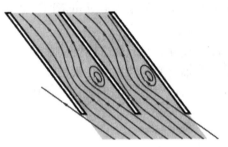

图 13.7 转轮或导叶入口处的冲击损失

13.3.6 圆盘摩擦

圆盘摩擦是指叶轮盖板和轮毂在充满液体的泵壳内旋转时造成的功率消耗的增加。叶轮和泵壳之间空腔内的液体开始旋转并形成一个主涡。主涡的旋转速度，在叶轮表面与叶轮相同，而在泵壳表面为零。因此假定主涡的平均速度为旋转速度的一半。

叶轮表面和泵壳表面液体的旋转速度不同产生的离心力将形成一个二次涡运动，如图13.8所示。由于二次涡把叶轮表面的能量传递给泵壳表面，所以二次涡将增加圆盘摩擦。

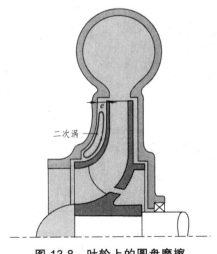

图 13.8 叶轮上的圆盘摩擦

圆盘摩擦的大小主要取决于转速、叶轮直径以及泵壳的尺寸，尤其是叶轮和泵壳之间的距离。此外，叶轮和泵壳表面的粗糙度对圆盘摩擦的大小也起着决定性的作用。叶轮外表面上的凸起或凹陷，如平衡块或平衡孔，也会增加圆盘摩擦。

13.3.7 渗漏

泵内旋转部件和固定部件之间的间隙处产生的回流会形成渗漏损失。与整个泵内的流量相比，叶轮内的流量增加了，所以渗漏损失会造成效率损失。

$$Q_{叶轮} = Q + Q_{渗漏} \tag{13.2}$$

式中，$Q_{叶轮}$为叶轮内的流量（立方米每秒），Q为泵内的流量（立方米每秒），$Q_{渗漏}$为渗漏量（立方米每秒）。

泵内很多地方都会产生渗漏，不同的泵渗漏处也不同。图13.9中给出了典型的渗漏处。图13.10中给出了泵内驱动渗漏流的压差。

通常叶轮入口处与轴向隙角处叶轮与泵壳之间的渗漏量是相同的。多级泵中，由于压差和间隙面积两者都很小，导叶和轴之间的渗漏量不太重要。

为了尽量减小渗漏量，间隙应越小越好。当间隙前后的压差很大时，缩小间隙则尤为重要。

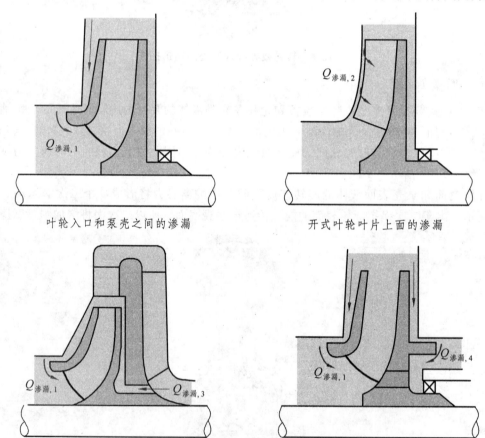

叶轮入口和泵壳之间的渗漏　　　　开式叶轮叶片上面的渗漏

多级泵中导叶与轴之间的渗漏　　　　平衡孔造成的渗漏

图13.9　渗漏的类型

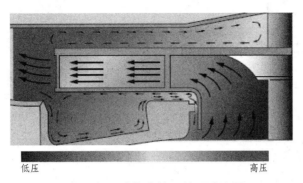

图 13.10 叶轮内的压差驱动渗漏

13.4 以比转速为函数的损失分布

前面所述的机械损失和水力损失的比例取决于比转速 n_s，比转速描述的是叶轮的形状。图 13.11 给出了设计工况点上损失的分布（路德维格等，2002）。

对于所有的比转速来说，流动摩擦与混合损失都是很重要的，对于高比转速而言，这两种损失是占主导地位的损失（半轴向和轴向叶轮）。对于低比转速的泵（径向叶轮），轮毂和盖板上的渗漏和圆盘摩擦一般会产生相当大的损失。

在非设计工况运行时，会产生冲击损失和回流损失。

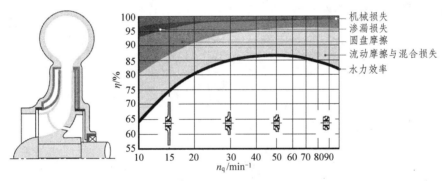

图 13.11 以比转速 n_s 为函数的离心泵中的损失分布

14 离心泵及其系统的特性曲线

离心泵是流体系统中最常见的组成部分。为了了解流体系统中离心泵的工作原理，有必要了解离心泵扬程与流量的关系。

14.1 泵的特性曲线

14.1.1 理论特性曲线

在假设泵的叶轮进口速度没有旋转分量的前提下，叶轮对单位重量的流体所做的功可用公式 14.1 进行计算。

$$\text{作用在单位重量流体上的功} = v_{w2}U_2/g \tag{14.1}$$

假设流体没有摩阻损失，泵的扬程可以看作与理论扬程相同。因此，可以把理论扬程写成

$$H_{\text{theo}} = \frac{v_{w2}U_2}{g} \tag{14.2}$$

从图 14.1 中的出口速度三角形可知，

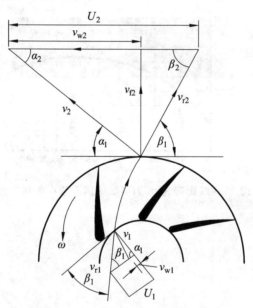

图 14.1 离心泵叶轮的速度三角形

$$v_{w2} = U_2 - v_{f2}\cot\beta_2 = U_2 - (Q/A)\cot\beta_2 \tag{14.3}$$

式中 Q ——叶轮出口的流量；
　　　A ——叶轮周围的过流面积。

出口的叶片转速 U_2 可以用叶轮转速 n 表示，$U_2 = \pi D n$。

根据这个公式以及公式 14.3，可以把公式 14.2 中的理论扬程改写成

$$H_{\text{theo}} = \pi^2 D^2 n^2 - \left[\frac{\pi D n}{A}\cot\beta_2\right]Q = K_1 - K_2 Q \tag{14.4}$$

式中，$K_1 = \pi^2 D^2 n^2$，$K_2 = (\pi D n / A)\cot\beta_2$。

对于一个转速恒定的叶轮，K_1 和 K_2 都是常数，因此扬程和流量之间呈线性关系，如公式 14.4。理论扬程 H_{theo} 随流量 Q 的线性变化如图 14.2 中的 I 所示。

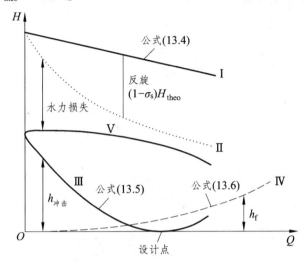

图 14.2 离心泵的扬程-流量曲线

14.1.2 实测特性曲线

流体流过叶轮的叶槽时，叶轮叶片迎水面和背水面的压力和流速是不同的。叶片迎水面压力相对较高、流速相对较低；而在背水面，压力较低、流速较大。导致在叶片周围产生环流，同一半径上的速度分布也不均匀。通常流体离开叶轮的角度小于实际叶片安放角。这种现象被称为"滑移（反旋）"。滑移系数（反旋系数）σ_S 是指叶轮出口实际的切线速度分量与理想的切线速度分量的差值。

如果考虑滑移的影响，理论扬程将降为 $\sigma_S v_{w2} U_2 / g$。而且滑移随着流量 Q 的增加而增大。通过图 14.2 中的曲线 II 可以看出滑移对扬程-流量曲线的影响。滑移产生的损失在真实流体和理想流体内都存在，但是在真实流体中还需要考虑叶片进口的冲击损失以及流道内的摩擦损失。

在设计点上，流体沿着切向方向进入叶片，所以冲击损失为零，但是除了设计点之外的其他点，冲击造成的水头损失按下面的关系增加

$$h_{\text{shock}} = K_3 (Q_f - Q)^2 \tag{14.5}$$

式中　Q_f ——非设计工况点的流量；

　　　K_3 ——常数。

摩擦造成的损失可以表示为

$$h_f = K_4 Q^2 \tag{14.6}$$

式中　K_4 ——常数。

图 14.2 中离心泵损失特性曲线Ⅲ和Ⅳ表示的就是公式（14.5）和（14.6）的损失。考虑滑移损失后，再在扬程曲线中减去任一流量下所有的损失之和（针对横坐标上所有点，从曲线Ⅱ的纵坐标中减去曲线Ⅲ和Ⅳ的纵坐标之和），就可以得到代表实际扬程的曲线Ⅴ，即为泵的实测扬程-流量特性曲线。

转速恒定时，离心泵的扬程 H、吸收功率 P、效率 η 和 $NPSH_R$ 都是流量 Q 的函数。这些参数值之间的关系即为特性曲线。一台转速 n=1450 转/分的单级离心泵的四个特性曲线如图 14.3 所示。

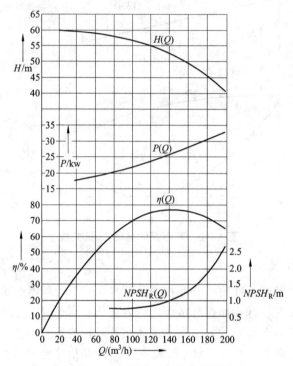

图 14.3　一台单级离心泵的特性曲线

扬程-流量 $H(Q)$ 曲线又称为节流曲线，代表一台离心泵的扬程与流量的关系。通常情况下，扬程随流量的增加而下降。一个扬程只对应一个流量。对扬程和流量的需求决定了泵的外形尺寸。

一台泵的吸收功率（功率消耗）曲线 $P(Q)$ 的形状也是流量的函数，如图 14.3 所示。对于径流泵，吸收功率随着流量的增加而上升，因此径流泵的启动通常采用闭闸启动的方式。吸收功率用来确定为泵提供能量的装置的尺寸。

效率曲线 $\eta(Q)$ 随着流量的增加从零开始增加，达到最大值（η_{opt}）后，开始下降。在选泵时除非考虑其他参数，否则所选泵的最高效率 η_{opt} 所对应的流量应尽量与系统所需流量 Q_r 相近，即 $Q_r \approx Q_{opt}$。效率曲线可以用来选择指定工作范围内效率最好的泵。

$NPSH_R$ 是必要气蚀余量的首字母简写。$NPSH_R$ 也是流量的函数。$NPSH_R$ 曲线在达到泵的高效点之前比较平缓，超过高效点后瞬速上升。为了避免泵发生气蚀，必须保证允许气蚀余量大于必要气蚀余量。允许气蚀余量（实际气蚀余量）是根据系统中的摩擦损失计算得到的，而必要气蚀余量是泵的供应商指定的。

14.2 系统的特性曲线

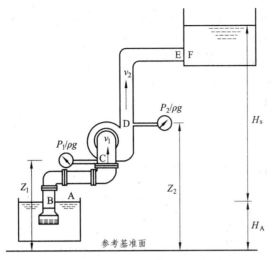

图 14.4 一般泵系统

下面考虑图 14.4 中的泵和管路系统。由于流体高度紊流，管路系统中的损失与流速的平方成正比，因此管路系统中的损失可以用恒定损失系数来描述。因此，吸水侧和压水侧的损失可以写成

$$h_1 = fl_1v_1^2/2gd_1 + K_1v_1^2/2g \tag{14.7a}$$

$$h_2 = fl_2v_2^2/2gd_2 + K_2v_2^2/2g \tag{14.7b}$$

式中　h_1——吸水侧的水头损失；

　　　h_2——压水侧的水头损失；

　　　f——达西摩擦系数；

　　　l_1, d_1 和 l_2, d_2——分别为吸水管和压水管的长度和直径；

　　　v_1 和 v_2——吸水管和压水管中的平均流速。

公式 14.7a 和 14.7b 中的第一项为普通摩擦损失（流体和管壁之间的摩擦损失，即沿程损失），第二项为损失系数 K_1 和 K_2 产生的所有较小的损失（局部损失），包括阀门、弯管、进口和出口损失等。因此泵把液体从低蓄水池抽送到高蓄水池所需提供的总扬程为

$$H = H_S + h_1 + h_2 \tag{14.8}$$

由于系统中的流量与流速成正比，因此以损失形式存在的流动阻力与流量的平方成正比，通常可以写成

$$h_1 + h_2 = 系统阻力 = KQ^2 \tag{14.9}$$

式中　K——常数，包括管道的长度和直径及各种损失系数。

公式 14.9 表示的系统损失是指任意给定流量流过系统时产生的水头损失。如果系统中任何参数发生变化，比如调整阀门开度或增加新的弯管等，K 值都会发生改变。因此公式 14.7 中的总扬程变为

$$H = H_S + KQ^2 \qquad (14.10)$$

扬程 H 可看做把流体从低蓄水池抽送到高蓄水池所需克服的管路中所有阻力水头之和。公式 14.9 为系统的特性方程，当画在 $H\text{-}Q$ 平面（图 14.5）上时代表的是系统特性曲线。

应该注意的是，如果液体没有净扬程的增加（如在同样高度的两个蓄水池之间的水平管道中抽水），则 H_S 等于零，系统扬程曲线将通过原点。

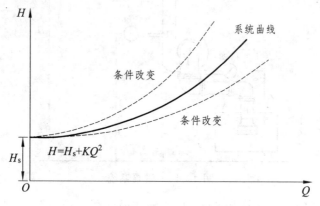

图 14.5 典型的系统水头损失曲线

14.3 泵特性与系统特性的匹配

水泵的设计点对应的是整体运行效率最高的情况。然而事实上水泵实际的工作点是通过把泵的水头损失-流量特性曲线与泵所接入的外部系统（如管路、阀门等）的特性曲线相匹配而得到的。把系统特性曲线与泵的特性曲线画在同一个坐标系中，就可以找到泵的工作点了。

$H\text{-}Q$ 平面内系统特性曲线与泵特性曲线的交点可能在也可能不在泵的最高效率点上，如图 14.6 所示。工作点与设计点的接近程度取决于对系统预期损失估计的准确性。工作点应位于优选的工作范围（POR）或允许工作范围（AOR）内。

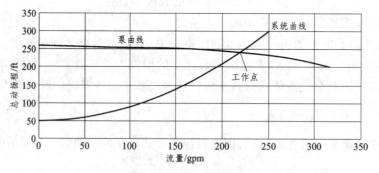

图 14.6 泵和系统的特性曲线

允许工作范围（AOR）是水泵制造商推荐的流量范围，如果泵在这个范围内工作，则泵的使用寿命不会严重受损。优选工作范围（POR）是客户指定的泵的最佳效率流量附近的一个范围。持续在优选工作范围内工作可以延长泵的使用寿命。POR 的推荐值为最高效率点流量的 70% 至 120% 之间。

15 离心泵在系统中的运行

本章介绍了离心泵在系统中的运行方式以及调节方式。离心泵总是与某一个系统相连的，在系统中使液体循环或提升。泵给流体增加能量用来克服管路系统中的摩擦损失或用来增加扬程。

把一台泵放在一个系统中，只会形成一个工况点。如果在同样情况下，几台泵联合运行时，把单个泵的扬程曲线串联或并联相加就可以得到该系统总的扬程曲线。调速泵可以通过调节转速来适应系统的需求。调速特别适用于加热系统（需热量取决于周围温度）和给水系统（需水量随用户开关龙头在不断变化）。

15.1 离心泵并联运行

对于那些流量变化很大而压力却相对恒定的系统，可以使两台或多台泵并联运行。多台泵运行时，可以同时调节一台或多台泵。为了避免液体在没有运行的泵内回流产生绕流循环，每台泵都串联连接了一个止回阀。图 15.1 为两台同型号的离心泵在相同的转速下并联工作。

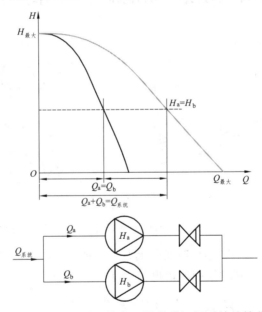

图 15.1 两台同型号的离心泵并联运行时的特性曲线

如图 15.1 所示，系统中每台泵的入口和出口都在同一点上，所以两台泵提供的扬程一定相同。然而，系统中的总流量是两台泵流量的总和。通过横向叠加每台泵的特性曲线，得到并联运行时的特性曲线，如图 15.1。

泵并联运行时，如果考虑系统特性曲线，两条曲线的交点即为工况点，该点的体积流量比单台泵的大，系统的水头损失比单台泵的高，如图 15.2 所示。这是由于体积流量的增加导致流速的增加，系统的水头损失也相应地增加。由于系统的扬程高，所以并联运行时实际的体积流量稍微小于单台泵流量的两倍。

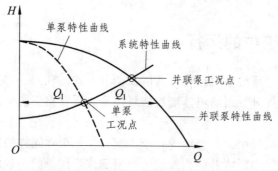

图 15.2 两台离心泵并联时的工况点

15.2 离心泵串联运行

离心泵串联运行是为了克服单台泵不能克服的更大的系统水头损失。如果串联运行的泵中有一台没有运行，则会给整个系统带来很大的阻力。为了避免上述情况，可以安装一个止回阀。如图 15.3 所示，两台同型号的离心泵在相同的转速下串联时，产生的体积流量相同，扬程也相同。由于第二台泵的进口与第一台泵的出口相连，所以两台泵产生的总扬程为两台泵扬程之和。从第一台泵进口到第二台泵出口的体积流量相同。

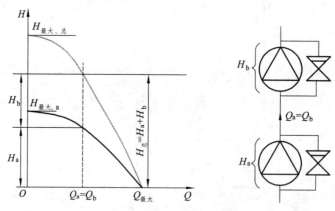

图 15.3 两台同型号的离心泵串联运行时的特性曲线

如图 15.4 所示，两台泵串联运行时实际上并没有把系统中的流动阻力加倍。两台泵为新的系统提供了足够的扬程，同时体积流量也稍微有所增加。

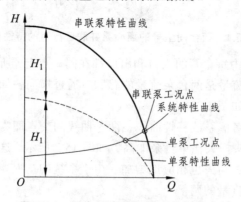

图 15.4 两台泵串联运行时的工况点

15.3 离心泵的工况调节

为某个特定系统选择泵时,保证工况点在泵的高效范围内是非常重要的。否则,泵的功率消耗将非常高。

然而,有时候选不到适应最优工况点的泵,因为系统的需求或系统特性曲线在不断地变化。因此,有必要调节泵的性能以满足需求的变化。

常见的泵性能调节方法有:

① 节流控制;
② 旁路控制;
③ 调速控制;
④ 叶轮切削。

根据对初期投资和泵运行成本的评估,选择合适的性能调节方法。除了叶轮切削之外,其他所有方法都可以在运行时连续的操作。通常情况下都为系统选择较大的泵,因此有必要限制其性能——首先是流速,有些时候也要限制最大扬程。

15.3.1 节流控制

为离心泵串联连接一个节流阀可以改变系统的特性,如图 15.5 所示。通过调整阀门的设置可以调节整个系统的阻力,进而达到所需的流量。系统特性曲线会变得更陡,与泵特性曲线的交点则会落在流量更低的点上。

节流阀会造成能量损失,因此连续使用节流阀进行控制会使效率降低。如果泵的特性曲线比较平缓,则可以使节流损失最小。因此节流控制主要用于径流泵的调节,因为径流泵的吸收功率随着流量的减小而降低。从控制系统的初期投资来看,采用节流阀进行控制有一定的优势,但是还应考虑整体经济状况,特别是装机功率高的情况。

对于混流泵和轴流泵来说,应注意吸入功率随着流量的降低而增加。此外,节流还可能使轴流泵进入不稳定的工作范围。这意味着运行不稳和噪音增加,两者都是高比转速泵的特点,所以在连续运行时应避免该工作区域。

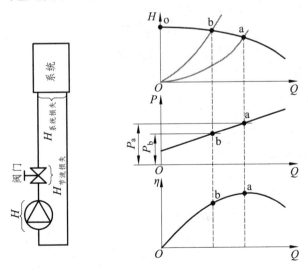

图 15.5 通过节流控制改变系统特性

15.3.2 旁路控制

旁通阀是与泵并联安装的调节阀,如图15.6所示。旁通阀使部分水流回流到吸水管线中,因此降低了扬程。安装旁通阀后,即使关断整个系统,泵还是会输送一定的流量。与节流阀类似,旁通阀在某些情况可以降低功率消耗。

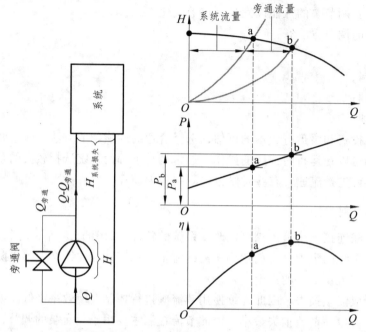

图 15.6 通过旁路控制改变系统特性

从整体来看,节流阀控制和旁通阀控制都不节能,应尽量避免。

15.3.3 调速控制

毫无疑问,在流量变化时,采用变频器的变速调节方法是效率最高的性能调节方法。该方法的优点在于这种调节直接减少输入系统的能量,而不是把多余的能量浪费掉。

当泵的转速改变后,相应的流量 Q、扬程 H 和功率 P 都会产生变化。转速的改变可以用相似定律来解释。这些定律为流量与转速成正比;扬程与转速的平方成正比;泵电机所需的功率与转速的立方成正比。以下公式对这些定律进行了总结。

$$Q \propto n, \quad H \propto n^2, \quad P \propto n^3 \tag{15.1}$$

式中　n——泵叶轮的转速(转/分);

　　　Q——泵的体积流量(加仑/分或立方英尺/小时);

　　　H——泵的扬程(磅/平方英寸或英尺);

　　　P——泵的功率(千瓦)。

采用这些比例率,可以根据某一转速下的特性,计算同一台泵不同转速下的特性。

$$Q_1\left(\frac{n_2}{n_1}\right) = Q_2, \quad H_1\left(\frac{n_2}{n_1}\right)^2 = H_2, \quad P_1\left(\frac{n_2}{n_1}\right)^3 = P_2 \tag{15.2}$$

即

$$\frac{H_2}{H_1}=\frac{Q_2^2}{Q_1^2} \quad 或 \quad H \propto Q^2 \tag{15.3}$$

由公式 15.3 可知，不同转速时扬程-流量特性曲线上的相应点或相似工况点都落在从原点出发的一条抛物线上。抛物线上的所有相似工况点都有相同的效率和比转速。所以这些抛物线又称为等效率曲线或等比转速曲线。

可以根据一台泵原转速时的特性曲线求出新转速时的特性曲线。方法是在原曲线上选几点，用相似定律来计算出新转速时的扬程和流量值。例如，在转速为 n_1 的特性曲线上取三点 A、B 和 C，见图 15.7。可以求出这三点在新的转速 n_2 时相应点 A'、B' 和 C'的扬程和流量值。

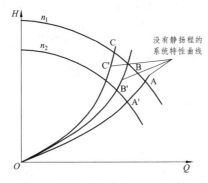

图 15.7 调速对离心泵工况点的影响

如果系统所需的性能发生变化，可以通过调节泵的转速来使泵在最优运行范围内工作。

15.3.4 叶轮切削

减小叶轮的直径是一个永久的变化，该方法可以在系统需求长期改变的情况下使用。叶轮切削后，如果更换电机降低能量消耗，则可以节能。可以通过相似定律来估计功率消耗、扬程和流量的变化，如图 15.8 所示。

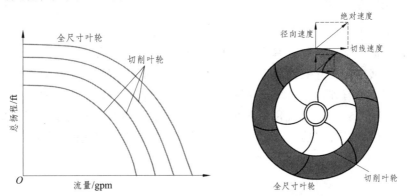

图 15.8 叶轮切削后的泵特性曲线

当转速恒定时，

$$\frac{Q_1}{Q_2}=\frac{D_1}{D_2}, \quad \frac{H_1}{H_2}=\left(\frac{D_1}{D_2}\right)^2, \quad \frac{P_1}{P_2}=\left(\frac{D_1}{D_2}\right)^3 \tag{15.4}$$

式中 D_1, D_2 ——叶轮切削前后叶轮的直径（毫米）；
Q_1, Q_2 ——叶轮切削前后泵的体积流量（英尺³/小时）；
H_1, H_2 ——叶轮切削前后泵的扬程（英尺）；
P_1, P_2 ——叶轮切削前后泵的功率（千瓦）。

切削是指采用机器加工的方法减小叶轮的直径。切削量应控制在原叶轮最大直径的20%之内，因为过度的切削会导致叶轮与泵壳不匹配。随着叶轮直径的减小，叶轮与固定泵壳之间的间隙会增大，从而增加内部回流，造成水头损失，降低泵的效率，如图15.9所示。

切削可以降低叶轮尖端的转速，从而降低了提供给流体的能量，因此泵的流量和压力都会下降。如果当前的叶轮产生的扬程过高，则可以采用较小的或切削的叶轮来提高效率。在实践中，叶轮切削通常用来避免控制阀门所产生的节流损失，切削后不会影响系统的流量。

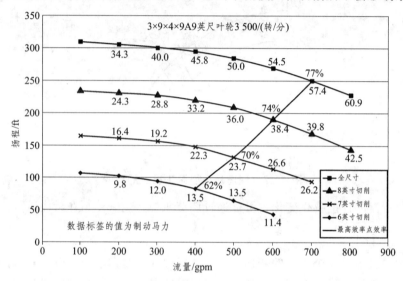

图 15.9　一台配有 9 英寸叶轮的离心泵几次切削之后的特性曲线

15.3.5　调节方法对比

阀门的使用使节流控制和旁路控制产生一些水力损失，因此降低整个系统的效率。叶轮切削在20%以内时不会对泵的效率产生很大的影响，因此这种方法不会对系统的总效率产生负面影响。只要转速不会降低到额定转速的50%以下，调速控制对泵的效率影响都很小。

每个方法都有利有弊，为系统选择调节方法时应把权衡所有利弊。如果要尽量保持最高的效率，最好采用叶轮切削方法和调速方法来降低流量。如果要求泵在固定的调整后的工况点运行，叶轮切削则是最好的方法。但是，对于流量需求不断变化的系统，采用调速泵是最好的解决办法。

16 离心泵的空化与气蚀余量

为了了解空化,首先必须了解蒸汽压。液体的蒸汽压是指液体在特定的温度下蒸发或转换为气体时的绝对压力。通常,蒸汽压的单位为磅/平方英寸(绝对压力)。液体的蒸汽压力随着温度的升高而增加,如图 16.1 所示。因此给定蒸汽压时必须指定温度。

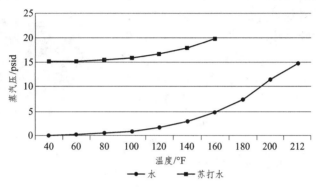

图 16.1 蒸汽压与温度的关系

在海平面上,水通常在 212 华氏度时沸腾。如果压力增加超过 14.7 磅/平方英寸,比如在锅炉或压力容器内,水的沸点也会增加。如果压力降低,水的沸点也会降低。例如,安第斯山脉的海拔为 15 000 英尺(4 600 米),正常大气压力为 8.3 磅/平方英寸而不是 14.7 磅/平方英寸,此时水的沸点仅为 184 华氏度。

泵内,由于流体流速的增加,叶轮入口的压力会降低。因此,液体可以在低压时沸腾。例如,如果叶轮入口处的绝对压力降到 1.0 磅/平方英寸,则水在 100 华氏度时就会沸腾或蒸发。

16.1 空 化

16.1.1 空化的定义

离心泵叶轮入口处的过流面积通常小于泵吸水管的过流面积或叶轮叶片的过流面积。当被抽送的液体进入离心泵叶轮入口时,过流面积的减小使流速增加,压力降低。流量越大,泵吸入口与叶轮入口之间的压降也越大。

如果压降足够大或温度足够高,当局部压力低于所抽送液体的饱和蒸汽压时,压降可能足以使液体蒸发。叶轮入口处压降形成的蒸汽气泡被流体带入叶轮叶片。当气泡远离叶片,进入局部压力大于饱和蒸汽压的区域时,气泡又会突然破灭。

如果气泡破灭的频率足够大,听起来就像玻璃球和石块在泵内流动。如果气泡破灭的能量足够大,则会移除泵壳内壁上的金属,留下像大的圆头锤打击过的凹痕,如图 16.2 所示。

图 16.2 空化的过程

16.1.2 空化的后果

离心泵中产生空化将严重影响泵的性能。空化会降低泵的性能和效率，并造成流量和出口压力的波动。由于液体被蒸汽所替代，所有流量会降低；叶轮流道中部充满了更轻的蒸汽，所以会造成机械不平衡。这将会导致振动和轴挠曲变形，最终导致轴承故障、填料或密封渗漏，甚至轴断裂。在多级泵中，空化会导致推力平衡损失或推力轴承故障。

空化还会破坏泵的内部组件。当泵产生空化时，在旋转叶轮叶片的正后面低压区会形成气泡。然后这些气泡向着迎面而来的叶轮叶片流动，并在叶片上面溃灭，对叶轮叶片的进水边（前缘）产生物理冲击。物理冲击会在叶片的进水边上击出小坑，内爆的气泡将移除叶轮表面的部分材料。每一个小坑的尺寸都是微小的，但是几小时或几天后，数百万计的这些小坑的累积效应则可以摧毁一个叶轮，如图 16.3 所示。

图 16.3 空蚀后的叶轮

仅有少数离心泵才允许在不可避免的空化条件下运行。但是必须对这些泵进行专门的设计和维护以抵抗运行过程中的少量空蚀作用。

16.1.3 空化的迹象

噪音和震动是离心泵产生空化的标志。如果泵在空化状态下运行时间过长，会出现以下情况：

- 出口压力、流量、泵电机电流产生波动；
- 叶轮叶片和泵壳内壁上形成蚀痕；
- 轴承过早被破坏；
- 泵轴断裂或其他疲劳破坏；
- 机械密封过早被破坏。

16.2 气蚀余量

为了避免离心泵中产生空化，泵内任何部位的流体压力都应高于饱和蒸汽压。用来确定被提升的液体压力是否足够避免空化的指标为净正吸入压头（气蚀余量）（NPSH）。气蚀余量是泵履行其职责所须达到的最低要求。因此，气蚀余量是指泵的吸入侧，包括叶轮入口处的情况。气蚀余量涉及吸水管道和连接部件、吸水管道中流体的高程和绝对压力，以及流体的流速和温度。简而言之，我们可以说气蚀余量决定了泵的进口要大于出口。

为了说明进入泵后流体的可用能量，NPSH 的度量单位采用英尺水头或泵吸水口的高程。允许气蚀余量 $NPSH_R$ 是针对泵而言的。实际气蚀余量 $NPSH_A$ 是针对系统而言的，系统是指泵吸水口侧的所有管道、水池和连接部件的整体。系统的 $NPSH_A$ 必须永远大于泵的 $NPSH_R$。

16.2.1 实际气蚀余量

实际气蚀余量（$NPSH_A$）是指泵进口压力与被提升液体的饱和蒸汽压之间的差值。$NPSH_A$ 是系统的特性，必须进行计算。$NPSH_A$ 是电站的设计人员根据提升液体的条件、泵的位置和高度、吸水管线的摩擦力等因素确定的。

$NPSH_A$ 的计算公式如下：$NPSH_A = h_A \pm h_Z - h_F + h_v - h_{vp}$

表 16.1 计算 $NPSH_A$ 时参数定义及备注

术语	定义	注释
h_A	供水箱中液体表面的绝对压力	・通常为大气压力（通气的水箱），但密闭的水箱不同 ・不要忘记高度对大气压的影响 ・总是正值（甚至可能更低，但即使是真空容器也是正绝对压力）
h_Z	供水箱中液体表面与泵中心线之间的垂直高差	・液位高于泵的中心线时为正值（称为静压头） ・液位低于泵的中心线时为负值（称为吸入压头） ・确保采用水箱中允许的最低水位
h_F	吸水管线中的摩擦损失	・液体向泵入口流动时，管道及配件会阻碍水流流动
h_v	泵入口处的速度头	・由于很小通常没有考虑
h_{vp}	提升温度下液体的饱和蒸汽压	・在最后要减去该值以确保入口压力高于蒸汽压力 ・需要注意的是温度升高，蒸汽压也升高

表 16.1 列出了计算实际气蚀余量时不同参数的定义和备注。图 16.4 为不同吸入条件的示意图。

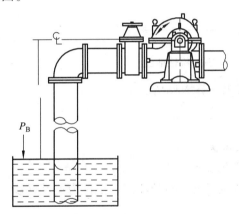

（a）带有吸升高度的开放式吸入条件

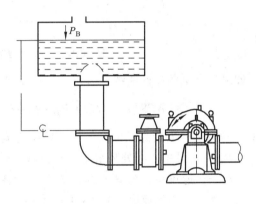

（b）带有吸入压头的开放式吸入条件

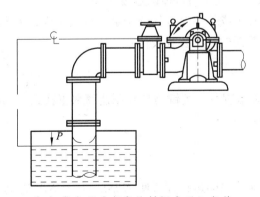

（c）带有吸升高度的封闭式吸入条件

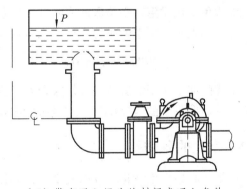

（d）带有吸入压头的封闭式吸入条件

图 16.4 不同吸入条件的示意图

16.2.2 允许气蚀余量

允许气蚀余量（$NPSH_R$）是在没有引起汽化（空化）的前提下，液体从泵入口流到叶轮入口所需克服的摩擦损失。$NPSH_R$是泵的特性，在泵的特性曲线上有标示。$NPSH_R$与很多因素有关，如叶轮入口的类型、叶轮设计、泵流量、叶轮转速，以及被提升液体的类型。$NPSH_R$是通过吸水高度测试确定的，形成一个以英尺汞柱为单位的负压后，转换成以英尺为单位的允许气蚀余量。

根据美国水力标准协会的规定，要对泵进行吸水高度测试，吸入容器中的压力要降到泵的总扬程损失达3%的点上。该点被称为泵的$NPSH_R$。一些泵的制造商通过关闭测试泵的进口阀门来进行类似的测试，另外一些制造商通过降低吸水高程来进行测试。

如果想知道所使用泵的$NPSH_R$值，最简单的方法就是从泵的特性曲线中读取。该值随着流量的变化而改变。泵文献中提到的$NPSH_R$值都是最高效率点的值。因此，我们很想知道泵运行时该曲线是什么样的。图16.5中给出了一条典型的$NPSH_R$特性曲线。

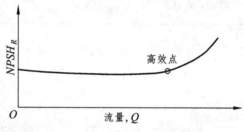

图 16.5 一条典型的 $NPSH_R$ 特性曲线

如果没有泵的特性曲线，可以通过以下公式来计算$NPSH_R$：

$$NPSH_R = ATM + P_{gs} + h_v - h_{vp}$$

式中 ATM——安装高程的大气压，单位为英尺水头；

P_{gs}——泵中心线处吸水压力计读数，转换为英尺水头；

h_v——速度头，$h_v = v^2/2g$，式中v为流体流过管道时的流速，单位为英尺每秒，g为重力加速度（32.16英尺每秒）；

h_{vp}——流体的蒸汽压，单位为英尺水头。蒸汽压与流体的温度有关。

要避免空化发生，必须保证实际气蚀余量大于或等于允许气蚀余量。也就是说，吸入端的压力必须大于泵所需的压力。该要求可以用下面的数学公式表达：

$$NPSH_A \geqslant NPSH_R$$

工程师都不希望泵安装完后，噪音大、缓慢或损坏。关键是从泵制造商那里找到$NPSH_R$的值，并确保系统的$NPSH_A$值大于泵的$NPSH_R$。

16.3 空化的预防

如果离心泵发生空化，可以改变系统的设计或泵的运行方式使$NPSH_A$大于$NPSH_R$,停止空化。增加$NPSH_A$的一种方法就是提高泵吸水口处的压力。例如，如果泵从一个密闭的水箱中抽水，可以提高水箱中的水位或增加液体上面的压力来提高吸入压力。

还可以通过降低被抽送液体的温度来增加 $NPSH_A$。因为降低液体的温度可以降低饱和压力,从而增加 $NPSH_A$。

如果可以降低泵吸水管线中的水头损失,则可以增加 $NPSH_A$。有很多种降低水头损失的方法,如增加管道直径,减少管道中弯管、阀门和配件的数量,以及减少管道的长度。

还可以通过降低泵的 $NPSH_R$ 来停止空化。在不同的运行状态下,给定泵的 $NPSH_R$ 并不是一个常数,而是随着一些因素而变化的值。通常情况下,随着泵流量的增加,泵的 $NPSH_R$ 明显增加。

因此,可以通过关小出水阀门来减小流量,从而降低 $NPSH_R$。$NPSH_R$ 还与泵的转速有关。泵叶轮的转速越快,$NPSH_R$ 越大。因此,如果可以降低调速泵的转速,也可以降低 $NPSH_R$。但是,由于泵的流量通常由泵所在系统的需求决定,在没有启动其他并联泵(如果有)的情况下,只能进行有限的调整。

17 轴流泵

轴流泵是三种离心泵的一种，另外两种为混流泵和径流泵。在三种离心泵中，轴流泵的特点是流量最大、出口压力最小。轴流泵可以提供的扬程为 10～20 英尺，远小于其他类型的离心泵。轴流泵可以处理很大的流量——可以达到数十万加仑每分钟，是离心泵中流量最大的一种泵。

轴流泵与径流泵的区别在于流体流入和流出轴流泵的方向都平行于转动轴。叶轮并不是给流体加速，而是把流体提升起来。叶轮的形状和运行原理都与船的螺旋桨类似，因此轴流泵也被称为旋桨泵。

17.1 轴流泵的基本构造

轴流泵是一种常见的泵，外形就像一根弯曲的水管，里面装有一个螺旋桨（轴流叶轮）。泵壳直径与吸入室的直径相差不大。轴流泵既可以垂直（立式）安装，又可以水平（卧式）安装，也可以倾斜（斜式）安装。轴流泵的基本部件有：吸入室、叶轮（包括叶片和轮毂）、导叶、轴、出水弯管、上下轴承、填料盒以及叶片角度的调节机构等，如图 17.1 所示。

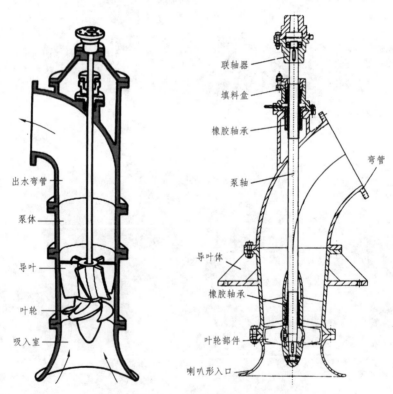

图 17.1 轴流泵结构图

1. 吸入室

为了改善轴流泵进口处的水力条件，一般采用喇叭形的进水口。

2. 叶轮

叶轮由中心轮毂和安装在轮毂上的一些叶片组成。叶轮上的叶片数量通常为 2～8 片，轮毂直径与叶轮直径比为 0.3 至 0.6。叶轮在圆柱形的泵壳内旋转，叶片叶梢与泵壳壁之间的间隙很小。叶轮可以由直接密封在泵壳内的电动机驱动，也可以由从侧面进入泵壳的驱动轴驱动。

叶片的安装方式使被提升的液体轴向流出（即与泵轴平行的方向），而不是径向（即与泵轴垂直的方向）。由于轴流泵的进口（吸入口）和出口（排出口）的半径变化很小，所以流体颗粒在流过泵的过程中没有改变其径向位置。在提升液体时，叶轮叶片的这种轴向布置产生的扬程很低。

叶轮叶片可以为固定式、半调式和全调式，如图 17.2 所示。固定式轴流泵的叶片与安装叶片的轮毂体是铸成一体的，叶片的安装角度不能调节。半调式轴流泵的叶片用螺栓栓紧在轮毂体上，可以调整叶片在轮毂体上的安装角度，但是必须要停机才能操作。全调式轴流泵，在任意时刻都可以通过一套油压调节机构来改变安装角度，进而使轴流泵在不同工况下高效运行。

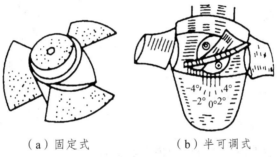

（a）固定式　　　　（b）半可调式

图 17.2　轴流泵叶轮

3. 导叶

出口处的导叶用来消除出口处流速的旋转分量，把旋转运动转换为压能。另外，还可以为轴流式水泵安装进口导叶，消除预旋转使水流完全轴向进入。

17.2　叶片设计

图 17.3 给出了一台轴流泵的简图。从图中可以看出水流流入和流出轴流泵的方向相同。图 17.4 为 X-X 剖面处叶片进口和出口的速度三角形。

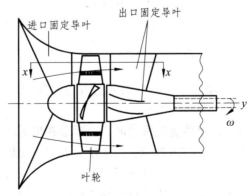

图 17.3　轴流泵的转轮

根据离心泵的欧拉方程，叶轮可以提供的扬程为

$$H = \frac{(U_2 \cdot C_{2U} - U_1 \cdot C_{1U})}{g} \quad (17.1)$$

式中，U 为切线速度，C_U 为绝对速度 C 在切线速度方向上的投影速度。

如果考虑平均半径 r_m 处的状态，则

$$U_2 = U_1 = U = r_m \cdot \omega \quad (17.2)$$

式中，ω 为叶轮的角速度。

叶轮向单位重量的液体所做的功为 $\dfrac{U \cdot (C_{2U} - C_{1U})}{g}$

为了实现最大的能量转换，$C_{1U} = 0$，即 $\alpha_1 = 90°$。另外，根据出口速度三角形，

$$C_{2U} = U - C_{2m} \cot \beta_2 \quad (17.4)$$

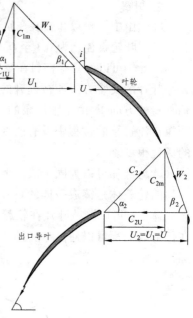

图 17.4 轴流泵的速度三角形

式中，C_m 为绝对速度 C 在径向的投影速度。

假设从进口到出口为恒定流动，

$$C_{1m} = C_{2m} = C_m \quad (17.5)$$

因此，轴流泵传递给单位重量液体的最大能量为

$$H = U \frac{(U - C_m \cot \beta_2)}{g} \quad (17.6)$$

由于不同半径处的质点获得的能量不同，流出叶轮后就会有能量交换，在能量交换过程中将损失很多能量。为了保证整个叶片上的能量转换恒定，以上方程对于所有的半径值 r 都应保持不变。但是，U^2 将随着半径 r 的增加而增加，为了保证上式恒定，必须保证 $UC_m \cot \beta_2$ 值同样增加。由于 C_m 是恒定的，因此 $\cot \beta_2$ 必须随着 r 的增加而增加。所以轴流泵的叶片随着半径的变化是扭曲的，如图 17.5 所示。

图 17.5 轴流泵扭曲的叶片

17.3 翼型理论

轴流泵的工作是以空气动力学中机翼的升力理论为基础的。叶轮上叶片的截面类似于与机翼的形状。

根据流体力学知识可知，流体流过机翼时，流体会在机翼上下分离成两股，它们分别经过机翼的上表面和下表面，如图 17.6 所示。由于沿机翼上表面的路程要比下表面的路程长一些，将造成机翼上面流速大、压力小，机翼下面流速小、压力大。流体对机翼的合力是向上的，这就是飞机升空的原理。同样，机翼对流体也产生一个向下的反作用力。

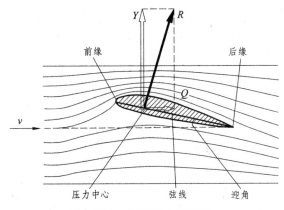

图 17.6 机翼升力示意图

Y—升力；R—总空气动力；Q—阻力。

在轴流泵中，翼型叶片倾斜固定在叶轮的轮毂上，如图 17.7 所示。翼型叶片随着轮毂的旋转而在流体中运动。翼型叶片的流线型恰好与机翼相反，使得翼型叶片上面流速小、压力大，翼型叶片下面流速大、压力小。流体对翼型叶片的合作用力是向下的，而翼型叶片对流体的作用力是向上的。在不断高速旋转的翼型叶片的作用下，液体被提升起来。

图 17.7 轴流泵叶轮的设计和流向

下面对与翼型（如图 17.8 所示）相关的术语进行定义：

· 吸力面（又称为上表面，轴流泵叶片背面）：通常流速大、静压力低。

· 压力面（又称为下表面，轴流泵叶片正面或工作面）：通常比吸力面上的静压力大。这两个表面之间的压力梯度为给定的翼型提供升力。

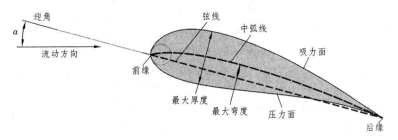

图 17.8 翼型命名法

用以下术语描述翼型的几何结构。

· 前缘（轴流泵叶片进水边）：是指翼型前面曲率最大（半径最小）的点。

· 后缘（轴流泵叶片出水边）：是指翼型后面曲率最大的点。

弦线：是指连接前缘和后缘的直线。弦长，简称弦，C 是指弦线的长度。弦长是翼型截面的参考尺寸。

用以下几何参数定义翼型的形状。

· 中弧线或等分线：为上表面和下表面中间点的轨迹。其形状取决于沿着弦线的厚度分布。

· 翼型的厚度：随弦线变化，可以采用下面任一种方式进行测量。

· 垂直于中弧线的厚度，有时被称为"美国惯例"；

· 垂直于弦线的厚度，有时被称为"英国惯例"。

最后是用来描述翼型在流体中移动时翼型行为的重要概念。

· 气动中心：仰俯力矩不受升力系数和迎角影响的弦向长度。

· 压力中心：仰俯力矩为零的弦向位置。

17.4 轴流泵的性能

轴流泵的工作特性与其他泵不同。制造商提供的轴流泵特性曲线用来说明单个泵扬程与流量的关系，该曲线也可以用来描述这些特性。图 17.9 为 Batescrew 公司制造的一台轴流泵的典型特性曲线。

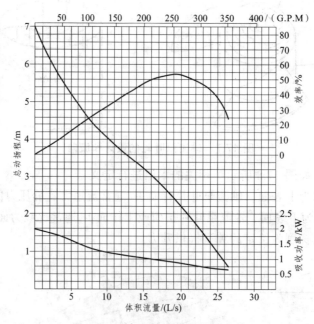

图 17.9 轴流泵的典型特性曲线

图中给出了轴流泵中扬程、流量、功率和效率之间的关系。虽然轴流泵在正常工作点上提供的扬程很低，但是其扬程随流量变化的曲线比其他种类的离心泵更陡。如图所示，轴流泵的关闭扬程（流量为零）是泵高效点扬程的三倍。另外，随着流量的减小功率需求增加，闭阀时所需能耗最大。这就是如果流量远远小于设计流量时，轴流泵的驱动电机过载的原因。这些趋势与径流式离心泵正好相反，径流式离心泵随着流量的增加，功率需求增加。

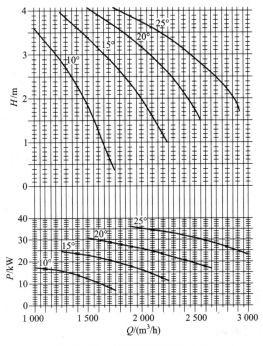

图 17.10 轴流泵的特性曲线

图 17.10 中的曲线为不同叶片节距（安放角）时轴流泵的性能变化情况。上面的线为叶片节距改变时扬程的变化，下面的线为吸收功率的变化。随着节距的增加，功率需求和扬程都增加，因此可以根据系统的条件调整叶片节距来达到最优的运行状态。

17.5 轴流泵的优点

轴流泵的主要优点就是在扬程相对较低时流量相对很大。例如，与常见的径流泵或离心泵相比，轴流泵在扬程小于 4 米时可提升三倍多的水或其他流体。通过改变桨叶节距，在小流量、高压力或大流量、低压力时，可以很容易的进行调节使轴流泵在高效范围内运转。

轴流泵中对流体的旋转作用不太大，叶轮叶片的长度也很短。这可以减小空气动力损失，提高分级效率。轴流泵的尺寸比许多传统泵都小，更适用于低扬程、大流量的情况。

17.6 轴流泵的应用

轴流泵适用于流量很大、压力（扬程）很低的情况。轴流泵的应用范围很广，如：
· 水库蓄水或泄洪；
· 洪水和暴风雨的控制；
· 电站或化工行业中大量水的循环；
· 大量的排水或灌溉；
· 原水取水；
· 大型城市污水处理厂中流体的输送；
· 沿海或低洼地区的水位控制；
· 干船坞和港口设施的蓄水或放空。

18 容积式泵

18.1 简 介

在容积式泵中,泵的每个工作循环都会输送一定体积的液体。该体积是一个常数,与泵所在系统中的流动阻力无关,前提是不能超过驱动泵的动力装置的容量或泵组件的强度极限。容积式泵输送的液体体积是相互独立的,两个相邻的体积之间没有流量,但是对于有多个腔室的泵,各个腔室的输送体积可能会重叠,可以减小这种效果。容积式泵与离心泵的区别在于离心泵在任何转速和流动阻力下输送的流量都是连续的。

根据设计和运行方式,可以把容积式泵分为三种基本类别:往复泵、转子泵和隔膜泵。

18.2 工作原理

所有容积式泵运行的基本原理都相同。用一台只有一个吸入口和一个排出口,且缸内装有单个往复式活塞的往复式容积泵,可以很容易的说明这个原理,如图 18.1 所示。吸入口和排出口上的止回阀使水流只能朝一个方向流动。

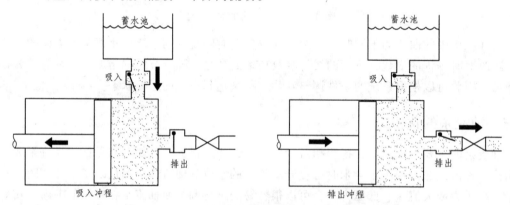

图 18.1 往复式容积泵的工作原理

在吸入冲程中,活塞向左移动,蓄水池和泵缸之间吸水管线上的止回阀打开,从蓄水池中吸水。

在排出冲程中,活塞向右移动,关闭吸水管线上的止回阀,打开排水管线上的止回阀。容积式泵单次循环(一个吸入冲程和一个排出冲程)输送的液体体积与活塞从最左侧移动到最右侧时泵缸内液体体积的改变量相同。

18.3 往复泵

一般把往复式容积泵分为四类:直接联动式或间接联动式,单缸式或双缸式,单作用式或双作用式,动力泵。

18.3.1 直接联动式和间接联动式往复泵

一些往复泵由原动机驱动,原动机也是往复式的,如往复蒸汽活塞驱动的往复泵。蒸汽活塞的活塞杆可以直接与泵的液体活塞相连，也可以通过梁或连杆间接相连。在直接联动式往复泵的液体（泵）端设有一个柱塞，柱塞直接由泵杆（及活塞杆或尾杆）驱动，柱塞携带动力端的活塞动作。间接联动式往复泵由梁或连杆驱动，该梁或连杆与一个独立的往复式发动机的动力活塞杆相连。

18.3.2 单缸式和双缸式往复泵

单缸式往复泵也被称为单缸泵，这种泵只有一个液（泵）缸。双缸式往复泵相当于在同一个基础上并排布置的两台单缸式往复泵。

双缸式往复泵两个活塞的驱动方式为，当一个活塞在上行冲程时，另一个活塞在下行冲程，反之亦然。与同类型设计的单缸泵相比，这种布置方式可以使双缸式往复泵的容量加倍。

18.3.3 单作用式和双作用式往复泵

单作用式（单动）往复泵只在一个方向吸水，即吸入冲程的方向为泵缸灌水，并在返回冲程，即排出冲程中把泵缸内的液体排出。在双作用式（双动）往复泵中，当液体从泵缸的一端吸入时，液体同时从泵缸的另一端排出。在返回冲程中，刚刚放空的泵缸一端充满液体，刚刚充满的一端放空。图18.2为单作用式和双作用式往复泵的一种布置方式。

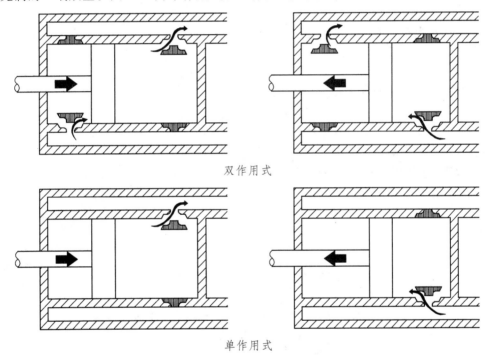

图 18.2 单作用式和双作用式往复泵

18.3.4 动力泵

动力泵通过减速齿轮、曲轴、连杆和十字头把旋转运动转换为低速往复运动。柱塞和活塞由十字头驱动器驱动。在低压高容量机组的液体端采用的杆和活塞构造类似于双缸双作用式蒸汽泵。高压机组一般为单作用式柱塞，通常采用三个柱塞。与单缸和双缸式往复泵相比，三个或多个柱塞可以明显减少流量的脉动。

动力泵通常具有较高的效率，并且可以形成很大的压力。动力泵可以用电动机或涡轮机驱动。动力泵是相对比较昂贵的一种泵，仅靠较高的效率是没法与离心泵相比的。然而，动力泵往往比蒸汽往复泵好，因为单作用式蒸汽泵对蒸汽的需求量很高，所以需要连续的运行。

一般来说，往复泵的有效流量随着被提升液体黏度的增加而减小，原因在于必须降低泵的转速。与离心泵相比，往复泵产生的压差与流体的密度无关。压差完全取决于施加在活塞上的力。

18.4 回转泵

回转泵的运行原理为旋转的叶片、螺杆或齿轮把液体储存在泵壳的吸入侧，然后迫使液体从泵壳的排出侧流出。由于回转泵可以把空气从吸水管线中排出，并且有较大的吸升高度，所以回转泵基本上都是自动引水的。对于需要较大吸水高度和具有自吸功能的泵来说，设计时必须保证旋转部件之间以及旋转部件与固定部件之间的所有间隙都最小，以降低回流效应。回流是指液体从泵的排出侧渗漏到泵的吸入侧。

由于回转泵的间隙很小，为了保证安全可靠的运行，并在较长的时间段内保证泵的流量不变，这些泵只能在相对较低的转速下运行。否则，高速液体流过狭窄的间隙时产生的腐蚀会很快引发过度磨损、间隙增大，从而导致回流。

往复式回转泵有很多种，通常可以分为三大类：齿轮泵、螺旋泵和动叶片泵。

18.4.1 简单齿轮泵

齿轮泵有几种不同的类型。图 18.3 中所示的简单齿轮泵由泵壳内沿着相反方向旋转的两个啮合在一起的正齿轮组成。泵壳与齿轮面及齿轮齿末端之间的间隙只有千分之几英寸。两个连续齿轮齿和泵壳之间的所有液体都必须跟随齿轮齿一起旋转。当齿轮齿与另一个齿轮齿啮合时，齿轮齿之间的空间减小，迫使夹带的液体从泵的排出管流出。齿轮继续旋转，齿轮齿分开，泵吸入侧的空间变大，吸入大量新的液体，并沿着泵壳把液体带到排出侧。随着液体从排出侧流出，形成一个低压，使液体从吸入侧流入。

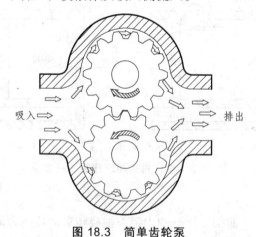

图 18.3 简单齿轮泵

齿轮上通常安装有很多齿轮齿，这样流量会相对比较平滑和连续，但是每次快速输送到排出管线的液体量较小。如果采用较少的齿轮齿，齿轮齿之间的空间变得更大，在给定的转速下,流量也更大,但是会增加脉动流量的倾向。在所有简单的齿轮泵中,功率都是施加到一个齿轮轴上的,然后通过齿轮齿啮合把功率输送给从动齿轮。

与往复泵不同的是,齿轮泵中没有造成摩擦损失的阀门。与离心泵不同的是,没有产生摩擦损失的高速旋转的叶轮。因此,齿轮泵适用于高黏性液体的输送,如燃料和润滑油。

18.4.2 容积式螺旋回转泵

容积式螺旋回转泵有很多不同的设计形式。主要的区别在于相互啮合的螺杆数、螺杆的螺距以及流体流动的方向。两种常见的类型为双螺旋低螺距双流量泵和三螺旋高螺距双流量泵。

1. 双螺旋低螺距螺旋泵

双螺旋低螺距螺旋泵由两个螺旋体组成,两个螺旋体分别安装在两个平行的轴上,螺旋体相互啮合时间隙很小。其中一个螺旋体为右旋螺纹,另一个螺旋体为左旋螺纹。其中一个轴为驱动轴,通过一组人字调速齿轮驱动另外一个轴。齿轮用来保证旋转螺旋体之间的间隙,并使其运行时更安静。两个螺旋体在紧密配合的双气缸内转动,气缸上设有重叠孔。所有的间隙都很小,但是两个螺旋体之间或螺旋体与泵缸壁之间是没有接触的。

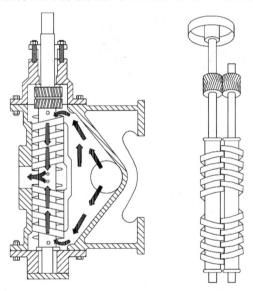

图 18.4 双螺旋低螺距螺旋泵

图 18.4 为双螺旋低螺距螺旋泵的完整组件和常见的流动路径。液体被困在每对螺旋叶片的外层空间。首先螺纹之间的第一个空间旋转远离相反的螺旋叶片,当螺旋叶片的末端再次与相反的螺旋叶片啮合时,将有一圈螺旋状的液体被困在该空间内。随着螺旋叶片继续旋转,夹带的螺旋状液体沿着泵缸滑向中部的排出空间,同时下一圈液体被带入。每个螺旋体的作用方式类似,由于每对螺旋叶片都沿着相反的方向向中部输送等量的液体,因此消除了液压推力。螺旋叶片把液体从吸入侧带走之后,会形成一个压力降,进而从吸入管线中吸入更多的液体。

2. 三螺旋高螺距螺旋泵

三螺旋高螺距螺旋泵与双螺旋低螺距螺旋泵有很多相同的部件,而且运行原理相同,如

图 18.5 所示。这种泵采用两端都有相反螺纹的三个螺旋体。三个螺旋体在一个三通缸内旋转，三通缸的两个外孔与中心孔重叠。螺旋体之间的螺距比低螺距螺旋泵要大很多，因此，中心的螺旋体或动力螺杆用来直接驱动两个外侧的从动螺杆，而不采用外部调速齿轮。底部的轴承支座用来支撑螺杆的重量，并保持他们的轴向位置。提升的液体进入吸入口，流过螺杆室周围的流道，然后以相反的方向流过两端的螺旋叶片，最后流向中部的排出口。不平衡的液压推力得以消除。螺旋泵可以用来提升黏性液体，如润滑油、液压油或燃油。

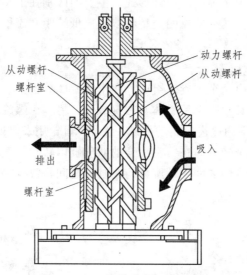

图 18.5　三螺旋高螺距螺旋泵

18.4.3　旋转式动叶片泵

图 18.6 所示的旋转式动叶片泵是另外一种常用的容积式泵。这种泵由开孔的圆柱形泵壳组成，吸入口在一侧，排出口在另一侧。泵缸内有一个直径比圆柱体略小的圆柱形转子，该转子由安装在泵缸中心线上方的轴进行驱动。转子与泵缸之间的间隙顶部小，底部大。转动时，转子带动叶片转进转出，保证转子与泵缸壁之间的密封空间。叶片从吸入侧引入液体或气体，并带向排出侧，然后空间的收缩使液体或气体从排出管线中排出。叶片可以在枢轴上摆动或在转子槽内滑动。

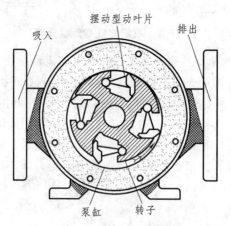

图 18.6　旋转式动叶片泵

18.5 隔膜泵

隔膜泵也被划分为容积式泵是因为隔膜类似于一个有限位移的活塞。机械联动装置、压缩空气或外源的脉动流体驱动隔膜做往复运动时，泵开始工作。泵的构造避免了被抽送的液体与能量源之间的任何接触。同时还消除了渗漏的可能性，在处理有毒或非常昂贵的液体时这一点是很重要的。这种泵的缺点在于有限的扬程和流量，以及吸入口和排出口中都要安装止回阀。图 18.7 为隔膜泵的实例。

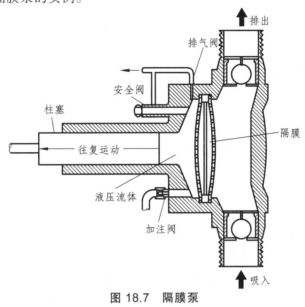

图 18.7　隔膜泵

18.6　容积式泵的特性曲线

容积式泵的每个工作循环都会输送一定体积的液体。因此，理想容积式泵的流量只受运行转速的影响。泵所在系统中的流动阻力不会影响泵的流量。

图 18.8 为容积式泵的特性曲线。图中的短划线为容积式泵的实际特性曲线。这条线反映了这样一个事实，随着泵出口压力的增加，一些液体会从泵的排出侧漏回吸入侧，从而降低泵的有效流量。液体从泵的排出侧漏回吸入侧的渗漏率被称为回流。

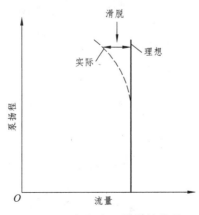

图 18.8　容积式泵的特性曲线

18.7　容积式泵的保护

为了防止泵及其排出管道压力过载,在容积式泵出口阀门的上游侧通常要安装安全阀。容积式泵可以为系统提供所需的压力。如果在运行时泵的出口阀关闭或出现过滤器堵塞限制系统流量等其他情况时,安全阀可以避免损坏系统和泵。

Glossaries (English to Chinese)

A

abrasive material　磨蚀物质；研磨材料
abscissa　横坐标
absent　缺少的
absolute pressure　绝对压力，绝对压强
absolute velocity　绝对速度
absorbed power　吸收功率
accelerate　加速
access door　人孔门；检修门；通道门
access shaft　入口井；竖井通道；进口竖井
access　*n.* 入口；通道；
　　　　v. 接近，进入；使用接近，获取
accessory　配件；附件
accommodate　容纳；使适应；供应；调解
accumulator　收集器
acidy water　酸性水
across　穿过；横穿
actuate　使动作
actuator　执行机构
acute angle　锐角
adhere　依附；坚持；粘着
adhesion　黏附；黏附力
adjust　调整，使……适合
adjustable blade Kaplan turbine　轴流转桨式水轮机
adjustable-blade　调节叶片
adjusting nut　调节螺母
adjustment ring　调整环
admit　进入
aerodynamic center　空气动力中心
aerodynamic force　空气动力
aerodynamic loss　空气动力损失
aerodynamics　空气动力学
aerofoil shaped　翼形；机翼的形状

affinity law　相似定律
aggressive　有腐蚀性；侵略性的；好斗的
air compensating valve　补气阀
air pressure　气压，风压
air valve　空气阀
air-bleed valve　排气阀
airfoil blade　翼型叶片
airfoil theory　翼形理论
airfoil　翼型；机翼
airtight　密封的
allowable operating range　允许工作范围
allowance　余量；限额
alter　改变，更改
alteration　改变；变更
alternative　比较方案
alternator　交流发电机
altitude　海拔；高度
ambient temperature　环境温度；周围温度
analogous　类似的
anchor bolt　锚固螺栓；地脚螺栓
anchor　使固定
Andes Mountains　安第斯山脉
angle of attack　攻角迎角；冲角；迎角；冲角
angle of inclination　倾角，倾斜角
angular frequency　角频率
angular velocity　角速度
annual production　年产量
annular area　环形区域
annular pipe　环形管
annular ring　环孔；环状垫圈
anticlockwise　逆时针
appreciable　相当可观的
arc　圆弧

area of flow 过流面积
arm 支臂
assembly 装配；组装
assume 承担；假定；采取；呈现
atmospheric pressure 大气压
attach 把……固定；贴上
automatic air valve 自动空气阀；自动排气阀
auxiliary servomotor 辅助接力器
available head 有效水头
axial bearing 止推轴承
axial direction 轴向
axial flow pump 轴流泵
axial flow turbine 轴流式水轮机
axial flow 轴流；轴向流动
axial force 轴向力
axial line 轴线
axial relief 轴向隙角
axial thrust 轴向推力
axis of rotation 旋转轴；转动轴
axis 轴
axle sleeve 轴套

B

babbit metal 巴比特合金
baffle 挡板
balance hole 平衡孔
balancing block 平衡块
balancing device 平衡装置
balancing hole 平衡孔
ball pein hammer 圆头铁锤
band 下环
Bánki-Michell turbine 双击式水轮机
base 基座；基础；底部
bearing bush 轴瓦
bearing housing 轴承箱；轴承体；轴承座
bearing loss 轴承损失
bearing pad 轴承垫片
bearing 轴承
behave 表现；运转
bend 弯管
Benoit Fourneyron 贝努瓦·富聂隆
BEP 高效点
Bernoulli's equation 伯努利方程
best efficiency point 高效点
bifurcate 分叉
blade angle 叶片安放角
blade arm 叶片臂
blade axis 叶片轴
blade channel 叶道
blade lever 叶片转臂
blade row 叶栅；叶片排
blade shape 叶片形状
blade 叶片
blade-operating mechanism 叶片操作机构
blow 打击
boil 煮沸，沸腾
boiler 锅炉；烧水壶，热水器
boiling point 沸点
bolt 螺栓；用螺栓固定
boss 轮毂；套筒
bottom ring 底环
bracket 支架；托架；括号
brake horsepower 制动功率
braking jet 制动射流
breakage 破坏；破损
bright-rolled 精轧的
bucket 斗叶、叶片
build-in 内置
bulb type tubular unit 灯泡式贯流机组
bulb 灯泡；灯泡体
burst 破裂
bypass circulation 旁路循环
bypass control 旁路控制
bypass valve 旁通阀
bypass 绕过；绕开；旁通管；旁通；旁路

C

cam disk 凸轮盘
cam 凸轮
camber line 弧线
cantilever 悬臂
capacity 流量；容量
capital cost 投资费用；资本成本
capsule 封装体
capture 捕获；捕捉
carbon steel 碳钢，碳素钢
casement 机壳
casing 机壳
cast steel 铸钢
cast wheel 铸轮
cast 浇铸
category 种类，分类
cavitate 形成空洞；成穴，空化
cavitation performance 空化性能
cavitation 空化；气穴现象；空穴作用
cavitational characteristic 空化特性
cavity seal 腔密封
cavity 空腔；洞
centerline 中心线
centrifugal compressor 离心式压缩机
centrifugal force 离心力
centrifugal pump 离心泵
chamber 腔，室
characteristic curve 特性曲线
check valve 止回阀
check 制止，抑制
chord line 弦线
circa 大约；近似
circulate 使循环
circumference 圆周；周长
civil cost 土建成本
civil work 土建工程；土建工作
clamp ring 锁紧圈
clear 避开；绕开

clearance 间隙；空隙
clockwise 顺时针
clog 阻碍；堵塞
clogged 阻塞的
closing device 关闭装置；闭合装置
closure 关闭；终止
coaxial 同轴的，共轴的
coaxially 同轴地
collapse 溃灭
compact structure 结构紧凑；密实结构
compensate 补偿
complication 并发症；复杂；复杂化
composition 构成；成分
compressor 压缩机
concentric 同轴的；同中心的
concrete 混凝土
configuration 配置；结构
conform 符合；遵照
conical hub 锥形轮毂
conical section 锥管段
conical 圆锥的；圆锥形的
connecting rod 连杆
connection plate 连接板
cons 缺点
constant efficiency curve 等效率曲线
constant specific speed curve 等比转速曲线
construction 构造；结构
container 容器；集装箱
contour 外形，轮廓
contraction 收缩，紧缩
control volume 控制体积；控制体
controlling mechanism 操作机构；控制机制
convert 转换
cooling 冷却
corner casing 护角铸件
correlation 相关，关联；相互关系
cosine 余弦

counter balance 平衡；配衡
couple 耦合
coupler 联轴器
coupling 联轴器；耦合
crankshaft 曲轴
criterion 标准
critical value 临界值
cross head 十字头；丁字头
cross section 横截面
crossflow water turbine 双击式水轮机
crosshead 十字头；联杆器
cross-sectional area 横截面积
crown 上冠
cube 立方
cumulative effect 累积效应
curvature 曲率
cut off 切断
cylinder 圆柱；圆柱体；油缸
cylinder-shaped runner 圆柱形转轮
cylindrical disc 圆柱形的轮辐
cylindrical 圆柱体的

D

dam 大坝
Darcy's friction factor 达西摩擦因子
dashed line 虚线
debris 碎片，残骸
decelerate 使减速
declared 宣布的
deflect 使转向；使偏斜；使弯曲
deflection 偏向；偏转；挠曲变形
deflector 偏流器；折向器
degrade 降低
deliver 输送
delivery 配送；传送，投递；交付
demand of water 需水量
denote 表示
dent 凹痕，凹部
depict 描述
Deriaz turbine 斜流式水轮机
derive 推导出；得到，导出
descend 下降
design parameter 设计参数；设计规范
designate 指定
destructive 破坏的；具有破坏性的
detachable 可拆式；可分开的；可拆开的
develop 产生；形成
deviation 偏离；偏差；误差
dewater 排水
diagonal turbine 斜流式水轮机
diagonal 斜的；对角线的
diaphragm pump 隔膜泵
dictate 控制，支配
diffuser 扩散器；导轮
dimension 把……刨成（或削成）所需尺寸
diminished 减少了的
direct-acting 直接联动式
discharge connection 压水管
discharge flow 排出流
discharge line 排水管线
discharge port 排出口
discharge rate 流量
discharge stroke 排出冲程
discharge tank 压水池
discharge n. 流量；
　　　　　 v. 排放；流出；放出
disconnection 断开；切断
disengage 分开
disk friction loss 圆盘摩擦损失
disk friction 圆盘摩擦
dismantle 拆开，拆卸
dismantling 拆开；拆卸
displace 取代；置换

disposition　排列；布置
dissipate　消散
distribution branch　配水支管
distribution pipe　配水管
distributor ring　控制环
distributor　进水管；配水管
diverging section　扩散段
divert　导向；使转移
divider　分水刃
dividing wall　隔墙；分隔墙
dome plate　顶板
dominant　占优势的；支配的
double acting jacking screw　双动式顶起螺丝
double acting　双动式
double volute pump　双蜗壳泵
double-acting　双作用式
double-regulated　双重调节
double-suction impeller　双吸式叶轮
double-suction　双吸
downstroke　下行程；下行冲程
draft tube cone　尾水锥

draft tube　尾水管
drain fitting　排水配件
drain valve　放空阀
drain　使流出；排掉
drainage　排水
drive horsepower　驱动功率
drive motor　驱动电动机
drive shaft　驱动轴；主动轴
drive　驱动
drive　驱动器
driving torque　驱动转矩；传动转矩；传动力矩
driving wheel　驱动轮；主动轮
dry dock　干船坞
dual control　双重控制
dump　扔弃
duplex　双缸式
duty point　工作点
dynamic action　动力作用；动态作用
dynamic head　动水头；动扬程

E

eddy current　涡流
eddy　涡流；漩涡
effective head　有效水头
efficiency curve　效率曲线
efficiency　效率
elbow discharge　出水弯管
elbow shape　肘形
elbow-shaped　肘形
electric motor　电动机
electric power　电力
electrical equipment　电气设备；电力设备
electrical generator　发电机
electrical load　电力负荷；电力负载
electrical output　电力输出
electrohydraulic　电动液压的

electronics　电子器件
elevation　高程；高度，海拔
elongation　伸长；延长
embed　嵌入；埋入
empirical　经验的
enclose　围绕；装入
enclosed impeller　封闭式叶轮
enclosed　封闭的
energy conversion　能量转换
energy dissipater　消能器
energy efficient　节能的
energy equation　能量方程
English system　英制
English unit　英制单位
enthalpy　焓

283

entrap 夹带
equilibrium equation 平衡方程
equilibrium 平衡
equivalent 等价的，相等的
erect 安装，建造
erosion 侵蚀，腐蚀
Euler 欧拉
Euler's equation 欧拉方程
even out 使均等
even 平坦的；相等的
ever-increasing 不断增加的
excavation 挖掘；开挖

exceptional 异常的，例外的
exciter 励磁机
execute 实行；执行
exert 施加（影响、压力等）
exordium 绪论
expand 扩张；扩大
expel 排出；驱逐；开除
expose 使暴露；使显露
expression 表达式
external diameter 外径
extract 提取；取出；榨取
extremity 末端；端点

F

fabricate 制造；装配
factory test 工厂测试
fan blade 风扇叶片；风机叶片
fan 风机
fastener 紧固件；扣件
feather 使……与……平行
feed 向……提供；流入，注入；喂养
feedback lever 反馈杆；反馈杠杆
feedback mechanism 反馈机制，回馈机制
feedback signal 反馈信号
feet 英尺
field fabrication 现场制作
filling opening 注油孔
finite 有限的
fin-shaped 鳍状；鳍形
fitting 配件；装置
fixed blade propeller turbine 轴流定桨式水轮机
fixed guide vane 固定导叶
fixed-blade 固定叶片
flange 法兰
flared 喇叭形；向外展开的
flash 闪蒸
flat efficiency curve 平坦的效率曲线

flexural stress 挠曲应力
float box 浮动框
float switch 浮动开关
floating 不固定的，流动的，浮动的
flow angle 进水角
flow area 过水面积；流动面积
flow friction loss 流动摩擦损失
flow inversion 流动逆转
flow path 流动路径
flow rate 流量
flow velocity 流速
fluctuate 波动；涨落
fluctuation 起伏，波动
fluid dynamics 流体动力学
fluid particle 流体质点；流体粒子
fluid 液体，流体
flush out 冲掉，排出
foreign object 异物；外物
forged 锻的；锻造的
fork U字形辊架
foundation 基础；地基
framework 框架，骨架；结构
Francis turbine 混流式水轮机；弗朗西斯式水轮机

frequency converter 变频器
frequency 频率
friction head 摩擦头；摩擦水头
friction joint 摩擦接头
friction loss 摩擦损失
friction 摩擦
frictional loss 摩擦损失

frictionless 无摩擦的；光滑的
from instant to instant 时时刻刻
fuel 燃料
full admission turbine 整周进水式水轮机
full gate opening 导叶全开度
fully tubular turbine 全贯流式水轮机
function 函数

G

gain 获得；增加
gallons per minute 加仑每分钟
gap 间隙；缺口
gas turbine 燃气轮机
gauge 计量器
gear face 齿轮端面
gear pump 齿轮泵
gear teeth 齿轮齿
gear 齿轮
generating unit 发电机组
generator hatch 发电机舱
generator rating 发电机额定值
generator rotor 发电机转子
generator set 发电机组
generator 发电机
geometrically 几何学上地
geometry edge 几何边缘；几何边界
geometry 几何形状；几何结构
gland packing 压盖填料
gland 压盖
glide over 滑过
govern 调节
governing device 调节装置

governing mechanism 调节机构；调速机构
governing 调速；控制，调节
governor mechanism 调速机构；调节器
governor 调速器
gpm 加仑每分钟
gravitation 重力
grease 润滑油；油脂
grid 电网
grind 磨碎；碾碎
grip 抓力；紧握
guide bearing 导轴承
guide passage 引水道，引流部件
guide ring 配水环
guide vane arm 导叶臂
guide vane cascade 导叶叶栅
guide vane mechanism 导叶操作机构；导水机构
guide vane 导叶
guide vanes mechanism 导水机构
guide vanes operating mechanism 导叶操作机构
guideline 指导方针；指导原则

H

half tubular turbine 半贯流式水轮机
halve 二等分
hamster cage 仓鼠笼
harbor installation 港口设施；海港设备

hardened 硬化的；淬火的
harness 利用
head cover 顶盖
head curve 扬程曲线

head loss   扬程损失；水头损失
head   水头；扬程
head-discharge characteristic curve   扬程-流量特性曲线
headwater   上游水位；上游源头
heat energy   热能
hemi-ellipsoided cup   半椭圆的杯子
herringbone   人字形的
hollow   空的；中空的
homologically   同源的
horizontal axis   水平轴，横轴；卧轴
horizontal pump   卧式泵
horizontal shaft   水平轴；卧轴
horizontal   水平的
horizontally   水平地

horsepower   马力
hose   软管
housing   机壳；外罩
hub   轮毂；中心
hydraulic efficiency   水力效率
hydraulic energy   水能
hydraulic head   水头，液压压头
hydraulic horsepower   水力功率
hydraulic loss   水力损失
hydraulic power   水力；水能；液压动力
hydraulic thrust   水推力；液压推力
hydraulic turbine   水轮机
hydrocarbon   碳氢化合物
hydroelectric energy   水电能源
hyperboloid   双曲面；双曲线体

I

identical   完全相同的
identically   同一地；相等地
identify   辨认；识别
idle position   停止位置
idler   惰轮；中间齿轮
illustration   说明；插图；例证；图解
impact   冲击，撞击
impart   传递
impeller diameter   叶轮直径
impeller eye   叶轮入口
impeller passage   叶轮流道；叶槽
impeller trimming   叶轮切削
impeller   叶轮
imperative   必要的，不可避免的
impinge   冲击；撞击
implode   向内破裂；内爆
implosion   向内破裂；内爆
improve   改进；改善，改良；增进
impulse turbine   冲击式水轮机
impulse   冲击式
in coordination with   配合；与……协调

in parallel   并联
in series   串联
in turn   依次；轮流地
inch   英寸
incidence loss   冲击损失
inclination   倾角
incline   使倾斜；使倾向于
incoming water   来水，进水
incur   招致，引发
indent   凹痕
indication   迹象；象征
indirect-acting   间接联动式
inertia   惯性；惰性
infinite   无限的，无穷的；无数的
inherent loss   固有损耗
initial investment   期初投资；最初投资
inject   注入；注射
injector   喷射机构；喷管
inlet area   进口面积
inlet diameter   进口直径
inlet guide vane   进口导叶

inlet rotation 入口旋转；入口回旋；进口旋转
inlet section 进口段；进口部分
inner face 内表面
innovative 创新的
inspection gate 检查口；检测门
inspection 检查；检验；检修
instability 不稳定（性）
installation 安装；装置
instantaneous load condition 瞬时负荷条件
intake 取水（口）
integrity 完整性

intercept 拦截，拦住
interfere 干涉；妨碍
intermediate transitory section 中间过渡段
intermediate 中间的
intermeshing 相互啮合的
internal diameter 内径
interpret 解释
intersection 交叉点
interstage bush 级间衬套
irrigation 灌溉

J

jacking screw 顶起螺丝；顶起螺钉；螺旋千斤顶
jam 使堵塞；挤进，使塞满

James B. Francis 詹姆斯·比切诺·法兰西斯
jet force 水冲力；喷射力
jet 射流

K

Kaplan turbine 卡普兰水轮机；转桨式水轮机
kilowatt 千瓦特，千瓦

kinetic energy 动能
kW 千瓦特，千瓦

L

lab test 实验室测试
labyrinth ring 迷宫环；曲折密封圈
lantern ring 水封环；套环
layout 布置方式；布局，安排，设计
leading edge 前缘
leading face 迎水面；工作面；正面
leak 使渗漏，泄露
leakage loss 渗漏损失；容积损失
leakage oil 渗漏油
leakage 渗漏；泄漏；渗漏物；漏出量
left-handed thread 左旋螺纹
Lester A. Pelton 莱斯特·阿伦·佩尔顿
level float 油位浮标
lever 转臂；操作杆；杠杆
lift test 提升测试
lift theory 提升理论

lift 提升
linear velocity 线速度
liner 衬垫，衬套；衬里
lining 衬里；内层；衬套
link ring 连接环
link support 支撑连杆
links and levers 连杆和转臂
literally 不夸张地
load limiter 载荷限制器；负荷限制器
load regulation 负荷调整
load rejection 甩负荷
load 负载，负荷
locking pin 定位销钉；锁定销
locus 轨迹
loss of energy 能量损失
loss 损失

low-lying area 低洼地区
lubricant 润滑油；润滑剂
lubricating oil 润滑油
lubrication 润滑

M

main circuit switch 主电路开关
main servomotor 主接力器
main shaft 主轴
main stay 主支柱
maintenance 维护
manhole 人孔，检修孔；探孔；入孔；进人孔
mantel 盖板；壁炉架
manufacturer 制造商
marble 大理石
mass flow rate 质量流量
mass flow 质量流量
maximum efficiency 最大效率
mean line 等分线
measurement scale 量尺；测量尺度
mechanic 技工，机修工
mechanical efficiency 机械效率
mechanical energy 机械能
mechanical loss 机械损失
mechanical seal 机械密封
mechanism 机制，机能，机理；机械装置
membrane 隔膜
mercury 汞，水银
mesh 啮合
metric 公制的；米制的
Michell-Banki turbine 双击式水轮机
microscopic 微小的；微观的

midriff 中腹部
mid-span 中跨
minus 减去
miscellaneous 各种各样的
mixed flow hydraulic turbine 混流式水轮机
mixed flow pump 混流泵
mixed flow turbine 混流式水轮机
mixed flow 混流
mixing loss 混合损失
model test 模型测试
moderate 适度的，中等的；温和的
modification 改进；变更；修改
molecule 分子；微小颗粒
moment of momentum equation 动量矩方程
momentum equation 动量方程
momentum 动量
motor horsepower 电机功率
motor 电动机
mount 安装
movable guide vane 活动导叶
movable vane 可动叶片
movable 可移动的；不固定的
moving vane pump 动叶片泵
multi-nozzle turbine 多喷嘴水轮机
multi-stage centrifugal pump 多级离心泵
multistage pump 多级泵
municipal 市政的

N

nameplate 铭牌
needle 喷针
negative head 负水头；负压头
negative pressure 负压
negative work 负功

net head 净水头
Net Positive Suction Head Available 允许气蚀余量
Net Positive Suction Head Required 必要气蚀余量

Net Positive Suction Head 气蚀余量
Newton's Law 牛顿定律
Newton's second law 牛顿第二定律
noise level 噪音水平
nominal efficiency 名义效率
nominal speed 额定转速
nominal 名义上的
non-return valve 止回阀
non-reversible 不可逆的
non-rotational 无旋的；非旋转的
nose cone 前锥体；泄水锥
notch 凹口；凹槽；缺口
nozzle 喷嘴
NPSH 气蚀余量
$NPSH_R$ 必要气蚀余量
nylon 尼龙

O

oblique division 倾斜的部分
obliquely 倾斜地
obviate 排除；避免；消除
occurrence 出现
off-design 非设计工况
offset 抵消；弥补
oil head 受油器
oil housing 集油罩
oil layer 油膜
oil level indicator 油位指示器
oil pipe 油管
oil piping 润滑油管系；油管
oil pressure system 油压系统
oil reservoir 储油器
oil scoop 集油盘
oil sling 油管吊索
oil tank 油箱
oil thrower ring 甩油环；甩油圈
oil transfer unit 受油器
onward 向前的
open impeller 敞开式叶轮
opening 开度；开口
operating cost 运营成本
operating mechanism 操作机构
operating member 运动部件
operating point 工况点
operating spindle 操作杆
operating time 操作时间；作业时间
optimum efficiency 最优效率
optimum 最适宜的
ordinate 纵坐标
oriented 确定……的方位
orifice 孔口
origin 原点
Ossberger turbine 双击式水轮机
outer edge 外缘；外刃
outer face 外面
outlet area 出口面积
outlet branch 出水支管
outlet diameter 出口直径
outlet guide vane 出口导叶
outlet tip 出水边
outlet velocity triangle 出口速度三角形
outlet 出口
output 出力
oval 椭圆形
overall efficiency 总体效率
overcome 克服；胜过
overhaul 大修；检修
overhead crane 桥式吊车，高架起重机
overhung 悬臂式的
overlap 重叠；重复
overlapping bore 重叠的孔
overload 超载；过载
overriding 过载
oversized 过大的，极大的

over-speed protective devices　超速保护装置
overspeed switch　超速开关

P

packing material　填充材料
packing ring　填料环；密封圈
packing　填料；包装；填充物
parabola　抛物线
parallel　平行；平行的；并联的
parallel-connected system　并联系统
parasitic loss　附加损失
part load　部分负荷
partial load　部分负荷；分载
partially　部分地
passage　通道
patent　*n*. 专利权；
　　　　　v. 授予专利；取得……的专利权
peak efficiency　最高效率
pedestal bearing　托架轴承；支承轴承
Pelton turbine　切击式水轮机；水斗式水轮机；佩尔顿水轮机
Pelton wheel turbine　佩尔顿式（切击式）水轮机
Pelton wheel　切击式水轮机
penetrate　穿透
penetration　渗透；突破；侵入
penstock　压力水管
perforated　穿孔的
performance curve　性能曲线
performance　性能；绩效
periphery　边缘；圆周；外围
permanent droop　永态转差率
perpendicular　垂直的，正交的
pilot servomotor　中间接力器
pilot valve　操纵阀
pipe wall　管壁
piston rod　活塞杆
piston servomotor　活塞接力器
piston　活塞

pit liner　基坑里衬
pit type tubular unit　竖井式贯流机组
pit　井；坑
pitch　节距；螺距；斜度，度
pitching moment　俯仰力矩
pitting mark　点蚀痕
pitting　点状腐蚀；点蚀
pivot　枢轴
plane　平面
plant factor　发电厂利用率；设备使用率
plant　电站
plunger　柱塞
pointer　指针
positive displacement pump　容积式泵；正排量泵
positive work　正功
potential energy　势能
pounds per square inch absolute　磅/平方英寸（绝对压强）
power consumption　功率消耗
power conversion　能量转化
power curve　功率曲线
power input　功率输入；电源输入
power output　功率输出
power plant　发电站；电厂
power pump　动力泵
power rating　额定功率
power unit　动力设备；动力装置
power　能量；动力；功率
powerhouse　厂房
precede　在……之前
preferred operating range　允许工作范围
premature　过早的；提前的
prescribe　规定；表示
prescribed　规定的

pressure action 压力作用
pressure drop 压力下降，压降
pressure energy 压能
pressure gauge 压力计，测压表
pressure gradient 压力梯度
pressure head 压力水头，压头，压位差
pressure side 迎水面；叶片正面；叶片工作面
pressure surface 压力面；叶片正面
pressure vessel 压力容器
pressurized water 加压水
prestressed 预应力
prevailing 占优势的；主要的；普遍的；盛行的
previous 先前的；稍前的
primary vortex 主涡
prime mover 原动机
product 乘积
profiled steel 异型钢
profiled 异形
progressively 渐进地；逐步的
project into 深入；插入
project 伸出；投影
projection velocity 投影速度
propeller pump 旋桨泵；轴流泵

propeller turbine 定桨式水轮机；螺旋桨（式）涡轮机
propeller 螺旋桨
proportional 比例的，成比例的
proportionalities 比例
pros 优点
protrude 使突出，使伸出
psia 磅/平方英寸（绝对压强）
pulsate 有规律地跳动
pulsation 脉动
pump casing 泵壳
pump characteristic curve 泵特性曲线；水泵性能曲线
pump input 水泵输入功率
pump output 水泵输出功率
pump shaft 泵轴
pump sump 废水收集池；集水坑
pump turbine 水泵水轮机
pump 泵
pumped storage plant 抽水蓄能电站
pumped storage 抽水蓄能
pumping head 抽送扬程
purge valve 冲洗阀
push-and-pull rod 推拉杆

R

racing 飞逸；空转
radial direction 径向
radial flow pump 径流泵
radial flow turbine 径流式水轮机
radial flow 径向流；辐流
radial-in-flow 径向流入式
radii 半径（radius 的复数）
radius 半径
rated output 额定输出；额定功率
raw water 原水
reaction turbine 反击式水轮机
reaction 反击式

Real pump curve 泵的实际曲线
recess 凹槽，凹处
reciprocating piston 往复活塞
reciprocating pump 往复泵；循环泵
recirculation loss 回流损失
recirculation zone 回流区
recover 回收；恢复；弥补
rectangular box 矩形盒子；矩形底座；矩形框架
reduction gearing 减速装置，减速齿轮
refill valve 加注阀
regulate 调节，规定；控制

regulating lever　调节杆；调整杆
regulating ring　控制环
regulating unit　调节装置
regulation valve　调节阀
regulator　调节器；调节装置；调控机构
rejection of load　甩负荷
relative velocity　相对速度
relief valve　安全阀；减压阀
remove　移除；移动，迁移
resemble　类似，像
resistance　阻力；电阻
resistant　有抵抗力的，抵抗的
resisting torque　阻力矩；抗力矩；抗转矩
resonant vibration　共振；谐振
resultant force　合力
resulting force　合力
resume　重新开始；恢复
retard　延迟；使减速
reveal　显示；透露；揭露
reverse　反向；倒转；反转
reversibility　可逆性
reversible hydraulic machinery　可逆式水力机械
reversible pump turbine　可逆式水泵水轮机
revert　恢复；使恢复原状；重提
revolutions per minute　转每分钟
revolve　旋转
rib　肋拱，肋材
right angle　直角

right-angle turn　直角转弯
right-handed thread　右旋螺纹
rigidly　刚性的；不易弯曲的
rim　边，边缘
ring area　环形区域
rise　高地；增加；上升
robust　结实的；耐用的；坚固的
rotary action　旋转运动
rotary pump　转子泵；旋转泵，回转泵
rotate　旋转
rotating element　转动元件，转动部分
rotating machinery　旋转机械
rotating shaft　转轴；旋转轴；回转轴
rotating　旋转；转动
rotation axis　旋转轴；转动轴
rotational speed　转速
rotor　转子；转轮
rough running　不平稳运转
rounding　使变圆；凑整
route　按某路线发送；给……规定路线
rub　摩擦
rubber ring　橡胶圈
runaway condition　失控工况；飞逸工况
runaway speed　飞逸转速
runaway　失去控制的
runner chamber　转轮室
runner vane　转轮叶片；轮叶
runner　转轮
run-of-the-river plant　河床式水电站

S

saddle point　鞍点
salient　显著的；突出的
saturation pressure　饱和压力
scalar quantity　标量
schematic view　示意图
screw pump　螺杆泵
screw stay　螺旋撑条；撑条

screw　螺杆
scroll case　蜗壳
scroll casing　蜗壳
sea level　海平面
seal joint　密封接头
seal surface　密封面；密封表面
sealing ring　密封环；密封圈；垫圈

sealing 封闭，密封
secondary vortex 二次涡
sediment collector 沉积收集器
seep out 渗出
segment 部分
self-aligning roller bearing 调心滚子轴承
self-priming 自灌式；自吸式
semicircular hood 半圆形罩
semi-open impeller 半开式叶轮
serve 具备
service life 使用寿命
servomotor piston 接力器活塞
servomotor 接力器；移动机构
servovalve 伺服阀
set about 着手；开始做…
setting angle 安装角；装置角
setting depth 埋深；设置深度
setting 安装；布置
severe 严重的；剧烈的
sewage treatment plant 污水处理厂
shaft extension type tubular unit 轴伸式贯流机组
shaft power 轴功率
shaft seal box 轴封盒
shaft seal device 轴封装置
shaft seal loss 轴封损失
shaft seal 轴封，轴封装置；主轴密封
shaft 轴
sharpen 削尖；磨快
shear pin 剪断销
shear stress 剪切应力
shield 盾；护罩；防护物
shock loss 冲击损失
shroud 盖板
shrouded impeller 闭式叶轮
shutdown 关机；停机
shut-off head 关闭扬程
shutoff valve 截流阀，截止阀
side stay 侧支柱

Siemens Martin steel 平炉钢
silt 淤泥，泥沙
simplex 单缸式
simplicity 简单；简易
simultaneously 同时地
single-acting 单作用式
single-stage pump 单级泵
single-suction impeller 单吸式叶轮
single-suction 单吸
site 现场；地点；位置；场所
slack 松弛的；不流畅的
slant 使倾斜；使倾向于
sleeve 套筒，套管
slide bearing 滑动轴承
slide block 滑块
slip factor 反旋系数
slip 反旋
slippage 回流
slope 斜率；倾斜；斜坡
slot 槽
slug 少量的液体
soda 苏打水；碳酸水
solid line 实线
span 跨度
spear rod 针阀
spear stem 喷杆
spear 喷杆
specific gravity 比重
specific speed 比转速
speed control 调速控制
speed-controlled pump 调速泵
spherical bearing 球面轴承
spider 支架；三脚架
spill 溢出，溅出
spin 快速旋转
spindle 轴
spiral casing 蜗壳
spiral-shaped 螺旋状
splash out 飞溅出来

split key 分半键
split runner 拼合式转轮
split volute pump 剖分式蜗壳泵
split 分开；分离；劈开
splitter 分水刃
spool position 阀芯位置
spool valve 柱形阀；短管阀
spoon-shaped 勺形；匙形
spring 弹簧
spur gear 正齿轮
square root 平方根
square 平方
squirrel cage turbine 鼠笼式水轮机
stage efficiency 分级效率
stagnation enthalpy 滞止焓
stainless steel 不锈钢
static discharge head 静排出压头，压水地形高度
static head 静压头；静水头；静扬程；落差
static pressure 静压
static suction head 静吸入压头；吸水地形高度
static suction lift 静吸上高度；静止吸入高度
stationary element 固定元件；固定部分
stay bolt 拉杆螺栓
stay cone 座环
stay ring 座环
stay shield 导流板
stay vane 固定导叶
stay vanes ring 固定导叶环
steady condition 稳定工况
steam piston 蒸汽活塞
steam turbine 蒸汽轮机
steel plate 钢板
steepness 倾斜度；陡度
steering arm 转臂
stiffen 使变硬；加强
stiffener 加固物；加劲杆；加强筋
straight flow type injector 直流式喷射机构，直喷嘴，内控式喷射机构
strainer 过滤器
strand 股（绳子的）
streamline 流线；流线型
streamlined 流线型的；改进的
strengthen 加强；巩固
strengthening rib 加强肋
strengthening spike 加强钉
stress 应力；压力
strike 冲击；打击
stroke 冲程
structural steel 结构钢，钢架
stuck 被卡住的；不能动的
stud 螺柱
stuffing box 填料盒；填料函
sturdy 坚固的
S-type turbine 轴伸式水轮机
subdivided 再分，细分
substitute 代替；替换
subtract 减去
subtype 子类型
succeeding 随后的
suction chamber 吸入室
suction connection 吸水管
suction flange 进口法兰
suction flow 吸入流
suction head 吸引高度；吸入水头
suction lift 吸水高度；吸升水头
suction line 吸水管线
suction nozzle 吸水口
suction piping 吸水管系统
suction port 吸入口
suction pressure 吸入压力
suction side 背水面；叶片背面；叶片负压侧
suction stroke 吸入冲程
suction surface 吸力面；叶片背面
suction tank 吸水池
support ring 支撑环
support stay 撑杆

surface roughness 表面粗糙度
surge protection 过载保护
susceptibility 易受影响或损害的状态；敏感性
swing 摆动
swivelable 可旋转的

symmetrical 对称的；匀称的
symmetry 对称（性）
synchronous speed 同步转速
system characteristic curve 系统特性曲线

T

tail race 尾水渠
tail water level 尾水位
tailrace channel 尾水渠
tailrace 尾水渠
tailwater elevation 尾水位
tailwater 尾水；下游水
tangent *adj.* 切线的；
  n. 切线，正切
tangential flow turbine 切向流式水轮机、冲击式水轮机
tangential velocity 切线速度；切向速度
tangential 切向的；切线的，正切的
tap 水龙头
taper 使成锥形；逐渐减少
teflon type fiber 聚四氟乙烯纤维
telescope connection 伸缩管式连接；伸缩节
temperature sensor 温度传感器
tensile stress 拉应力
terminate 结束，终止
theoretical foundation 理论基础
theoretical head 理论扬程
theoretical loss-free head 无损失理论扬程
theoretical pump curve 泵的理论曲线
Thoma cavitation coefficient 托马空化系数
thousandths 千分之一的
three-screw, high-pitch, screw pump 三螺旋高螺距螺旋泵
throat 喉口
throttle control 节流控制
throttle valve 节流阀
throttling curve 节流曲线

through-flow 过流
thrust bearing 推力轴承；止推轴承
thrust-cum-guide bearing 推导组合轴承
tidal power 潮汐能
tie 加强带；束缚
tilt 倾斜；翘起
timing gear 调速齿轮；正时齿轮
tip 尖端
toggle lever 曲柄连杆
torque pin 扭力销；转矩销
torque 扭矩
torsional stress 扭曲应力
total dynamic discharge head 总动排出压头
total dynamic head 总动压头
total dynamic suction head 总动吸入压头
total dynamic suction lift 总动吸上高度；总动吸入高度
total efficiency 总效率
total head 总水头；总压头
total static head 总静水头
trailing edge 后缘；机翼后缘
trailing face 背水面
transmit 传输；传送，传递
transversely 横着；横断地；横切地
trapezoidal 梯形的
trigger 引发，引起；触发
trimmed impeller 切削的叶轮
trimming 切削；修剪
trough-shaped blade 槽形叶片
trough-shaped 槽形的
trunnion 耳轴

tubular turbine 贯流式水轮机
turbine chamber 水轮机室
turbine governing 水轮机调速
turbine setting 水轮机安装高程
turbo 涡轮
turbomachine 透平机械；透平机械；涡轮机
turbulent 湍流的
Turgo turbine 斜击式水轮机；土戈尔式水轮机
twisted 扭曲的
two-screw, low-pitch, screw pump 双螺旋低螺距螺旋泵

U

unattended 无人值守的
uneconomical 不经济的；浪费的
unit head 单位水头
unit power 单位功率
unit 机组
unwieldy 笨拙的；笨重的
upstream face 上游面
upstroke 上行程；上行冲程
utilize 利用

V

vacuum 真空
vane 叶片
vaneless 无叶片的
vapor bubble 气泡
vapor pressure 蒸汽压
vaporization 蒸发
vaporize 使……蒸发；蒸发
variable pressure turbine 变压力水轮机
variation of speed 变速；改变转速
variation 变化，变动；变异；变种
vector addition 向量加法；矢量加法
velocity component 速度分量
velocity energy 速度能
velocity gradient 速度梯度
velocity head 速度头；动压头
velocity profile 速度（流速）剖面
velocity triangle 速度三角形
velocity vector 速度矢量
velocity 流速、速度
vent fitting 排气装置
vented 通风的；开孔的
versatile 多用途的；多功能的
vertical axis 垂直轴；纵轴；立轴
vertical pump 立式泵
vertical shaft 垂直轴；立轴
vertical 垂直的
vertical-axis 竖轴；垂直式
Viktor Kaplan 维克多·卡普兰
virtue 优点
viscosity 黏度
viscous fluid 黏性流体
volume 量；体积
volumetric efficiency 容积效率
volumetric flow rate 体积流量
volumetric 体积的；容积的
volute casing 蜗形机壳；蜗壳
volute tongue 蜗壳舌部；蜗壳隔舌
volute 蜗壳；螺旋形；涡形
vortex 涡流；漩涡

W

waste water 废水；污水
water column 水柱

water conduit 水管
water diversion part 引水部件
water drainage part 泄水部件
water energy 水能
water horsepower 水功率
water passage 流道
water tight casing 水封的机壳
water wheel 水轮机；水车
wear disk 耐磨盘
wear ring 耐磨环；磨损环；磨耗环
wear 磨损；耗损
wearing ring 承磨环；减磨环，抗磨环
web 盖板；腹板

webbing 盖板；边带
wedge 楔块
weld 焊接
wheel 轮毂；转轮
whirl component 旋转分量；旋转组件
whirl 旋转，回旋
whirling component 旋转分量
wicket gate 活动导叶
wind turbine 风轮机，风力涡轮机
windage loss 风阻损失
wing-shaped 翼状的
work 功
working member 工作部件

Y

yield 生产；屈服

yoke 轭架

Glossaries (Chinese to English)

A

安第斯山脉　Andes Mountains
安全阀　relief valve
安装　erect, install, mount, setting
安装角　setting angle
鞍点　saddle point
凹槽　notch
凹处　recess
凹痕　dent, indent

B

巴比特合金　babbit metal
摆动　swing
半贯流式水轮机　half tubular turbine
半径　radius, radii (plural)
半开式叶轮　semi-open impeller
半椭圆的杯子　hemi-ellipsoided cup
半圆形罩　semicircular hood
磅/平方英寸（绝对压强）　pounds per square inch absolute, psia
饱和压力　saturation pressure
暴露　expose
贝努瓦·富聂隆　Benoit Fourneyron
背水面　suction side, trailing face
笨重的　unwieldy
泵　pump
泵壳　pump casing
泵特性曲线　pump characteristic curve
泵轴　pump shaft
比较方案　alternative
比例　proportionalities
比重　specific gravity
比转速　specific speed
必要的　imperative
必要气蚀余量　Net Positive Suction Head Required, $NPSH_R$
闭合装置　closing device
闭式叶轮　shrouded impeller
壁炉架　mantel
避开　clear
避免　obviate
边缘　rim
变动　variation
变更　alteration, modification
变化　variation
变频器　frequency converter
变速　variation of speed
变压力水轮机　variable pressure turbine
辨认　identify
标量　scalar quantity
标准　criterion
表达式　expression
表面粗糙度　surface roughness
表示　denote
表现　behave
并发症　complication
并联　in parallel
并联的　parallel
并联系统　parallel-connected system
波动　fluctuate
伯努利方程　Bernoulli's equation
补偿　compensate
补气阀　air compensating valve
捕获　capture
布置　setting

捕捉　capture
不断增加的　ever-increasing
不固定的　floating
不经济的　uneconomical
不可避免的　imperative
不可逆的　non-reversible
不夸张地　literally
不平稳运转　rough running

不稳定（性）　instability
不锈钢　stainless steel
布局　layout
布置　disposition
布置方式　layout
部分　segment
部分地　partially
部分负荷　part load, partial load

C

仓鼠笼　hamster cage
操纵阀　pilot valve
操作杆　operating spindle
操作机构　controlling mechanism, operating mechanism
操作时间　operating time
槽　slot
槽形的　trough-shaped
槽形叶片　trough-shaped blade
侧支柱　side stay
测量尺度　measurement scale
测压表　pressure gauge
插入　project into
插图　illustration
拆开　dismantle
拆卸　dismantling
产生　develop
厂房　powerhouse
敞开式叶轮　open impeller
超速保护装置　over-speed protective devices
超速开关　overspeed switch
超载　overload
潮汐能　tidal power
沉积收集器　sediment collector
衬垫　liner
衬里　lining
衬套　liner
撑杆　support stay

成比例的　proportional
承担　assume
承磨环　wearing ring
乘积　product
齿轮　gear
齿轮泵　gear pump
齿轮齿　gear teeth
齿轮端面　gear face
冲程　stroke
冲掉　flush out
冲击　impact, impinge, strike
冲击式　impulse
冲击式水轮机　impulse turbine, tangential flow turbine
冲击损失　incidence loss, shock loss
冲洗阀　purge valve
抽水蓄能　pumped storage
抽水蓄能电站　pumped storage plant
抽送扬程　pumping head
出口　outlet
出口导叶　outlet guide vane
出口面积　outlet area
出口速度三角形　outlet velocity triangle
出口直径　outlet diameter
出力　output
出水边　outlet tip
出水弯管　elbow discharge
出水支管　outlet branch

出现　occurrence
储油器　oil reservoir
触发　trigger
穿过　across
穿孔的　perforated
穿透　penetrate
传递　impart
传动转矩　driving torque
传输　transmit

传送　delivery
传送　transmit
串联　in series
创新的　innovative
垂直的　perpendicular, vertical
垂直轴　vertical axis, vertical shaft
凑整　rounding
淬火的　hardened

D

达西摩擦因子　Darcy's friction factor
打击　blow, strike
大坝　dam
大理石　marble
大气压　atmospheric pressure
大修　overhaul
大约　circa
代替　substitute
单缸式　simplex
单级泵　single-stage pump
单位功率　unit power
单位水头　unit head
单吸　single-suction
单吸式叶轮　single-suction impeller
单作用式　single-acting
弹簧　spring
挡板　baffle
导流板　stay shield
导轮　diffuser
导水机构　guide vane mechanism, guide vanes mechanism
导叶　guide vane
导叶臂　guide vane arm
导叶操作机构　guide vane mechanism, guide vanes operating mechanism
导叶全开度　full gate opening
导叶叶栅　guide vane cascade

导轴承　guide bearing
倒转　reverse
灯泡　bulb
灯泡式贯流机组　bulb type tubular unit
等比转速曲线　constant specific speed curve
等分线　mean line
等价的　equivalent
等效率曲线　constant efficiency curve
低洼地区　low-lying area
抵抗的　resistant
抵消　offset
底环　bottom ring
地点　site
地基　foundation
地脚螺栓　anchor bolt
点蚀　pitting
点蚀痕　pitting mark
电厂　power plant
电动机　motor, electric motor
电动液压的　electrohydraulic
电机功率　motor horsepower
电力　electric power
电力负荷　electrical load
电力输出　electrical output
电气设备　electrical equipment
电网　grid
电站　plant

点状腐蚀　pitting
电子器件　electronics
丁字头　cross head
顶板　dome plate
顶盖　head cover
顶起螺钉　jacking screw
顶起螺丝　jacking screw
定桨式水轮机　propeller turbine
定位销钉　locking pin
动力　power
动力泵　power pump
动力设备　power unit
动力装置　power unit
动力作用　dynamic action
动量　momentum
动量方程　momentum equation
动量矩方程　moment of momentum equation
动能　kinetic energy
动水头　dynamic head

动态作用　dynamic action
动压头　velocity head
动扬程　dynamic head
动叶片泵　moving vane pump
斗叶　bucket
堵塞　clog
堵塞　jam
断开　disconnection
锻造的　forged
对称（性）　symmetry
对称的　symmetrical
对角线的　diagonal
盾　shield
多级泵　multistage pump
多级离心泵　multi-stage centrifugal pump
多喷嘴水轮机　multi-nozzle turbine
多用途的　versatile
惰轮　idler
惰性　inertia

E

额定功率　power rating, rated output
额定输出　rated output
额定转速　nominal speed
轭架　yoke

耳轴　trunnion
二次涡　secondary vortex
二等分　halve

F

发电厂利用率　plant factor
发电机　generator, electrical generator
发电机舱　generator hatch
发电机额定值　generator rating
发电机转子　generator rotor
发电机组　generating unit, generator set
发电站　power plant
阀芯位置　spool position
法兰　flange
反击式　reaction
反击式水轮机　reaction turbine

反馈杆　feedback lever
反馈机制　feedback mechanism
反馈信号　feedback signal
反向　reverse
反旋　slip
反旋系数　slip factor
放空阀　drain valve
飞溅出来　splash out
飞逸　racing
飞逸工况　runaway condition
飞逸转速　runaway speed

非设计工况	off-design		封装体	capsule
非旋转的	non-rotational		弗朗西斯式水轮机	Francis turbine
废水	waste water		浮动的	floating
沸点	boiling point		浮动开关	float switch
沸腾	boil		浮动框	float box
分半键	split key		符合	conform
分叉	bifurcate		辐流	radial flow
分隔墙	dividing wall		俯仰力矩	pitching moment
分级效率	stage efficiency		辅助接力器	auxiliary servomotor
分开	disengage, split		腐蚀	erosion
分离	split		负功	negative work
分水刃	divider, splitter		负荷	load
分子	molecule		负荷调整	load regulation
风机	fan		负荷限制器	load limiter
风机叶片	fan blade		负水头	negative head
风力涡轮机	wind turbine		负压	negative pressure
风轮机	wind turbine		负压头	negative head
风扇叶片	fan blade		负载	load
风阻损失	windage loss		附加损失	parasitic loss
封闭的	enclosed		腹板	web
封闭式叶轮	enclosed impeller			

G

改变	alter		隔墙	dividing wall
改变转速	variation of speed		各种各样的	miscellaneous
改进	improve		工厂测试	factory test
改善	improve		工况点	operating point
盖板	shroud, web, webbing		工作部件	working member
干船坞	dry dock		工作点	duty point
干涉	interfere		工作面	leading face
刚性的	rigidly		公制的	metric
钢板	steel plate		功	work
港口设施	harbor installation		功率	power
杠杆	lever		功率曲线	power curve
高程	elevation		功率输出	power output
高效点	best efficiency point, BEP		功率输入	power input
隔膜	membrane		功率消耗	power consumption
隔膜泵	diaphragm pump		攻角	angle of attack

汞　mercury
共振　resonant vibration
构成　composition
构造　construction
股（绳子的）　strand
骨架　framework
固定　attach
固定　anchor
固定导叶　fixed guide vane, stay vane
固定导叶环　stay vanes ring
固定叶片　fixed-blade
固定元件　stationary element
固有损耗　inherent loss
关闭　closure
关闭扬程　shut-off head
关闭装置　closing device
关机　shutdown
管壁　pipe wall

贯流式水轮机　tubular turbine
惯性　inertia
灌溉　irrigation
规定　prescribe
规定的　prescribed
规定路线　route
轨迹　locus
辊架　fork
锅炉　boiler
过大的　oversized
过流　through-flow
过流面积　area of flow
过滤器　strainer
过水面积　flow area
过载　overload, overriding
过载保护　surge protection
过早的　premature

H

海拔　elevation, altitude
海平面　sea level
函数　function
焓　enthalpy
焊接　weld
合力　resultant force, resulting force
河床式水电站　run-of-the-river plant
横断地　transversely
横截面　cross section
横截面积　cross-sectional area
横着　transversely
横轴　horizontal axis
横坐标　abscissa
喉口　throat
后缘　trailing edge
弧线　camber line
护角铸件　corner casing
护罩　shield

滑动轴承　slide bearing
滑过　glide over
滑块　slide block
环境温度　ambient temperature
环孔　annular ring
环形管　annular pipe
环形区域　annular area, ring area
恢复　recover, resume, revert
恢复原状　revert
回流　slippage
回流区　recirculation zone
回流损失　recirculation loss
回收　recover
回旋　whirl
回转泵　rotary pump
混合损失　mixing loss
混流　mixed flow
混流泵　mixed flow pump

混流式水轮机　Francis turbine, mixed flow turbine, mixed flow hydraulic turbine
混凝土　concrete
活动导叶　movable guide vane, wicket gate
活塞　piston
活塞杆　piston rod
活塞接力器　piston servomotor
获得　gain

J

机壳　casing, housing, casement
机理　mechanism
机械密封　mechanical seal
机械能　mechanical energy
机械损失　mechanical loss
机械效率　mechanical efficiency
机修工　mechanic
机制　mechanism
机组　unit
基础　foundation
基坑里衬　pit liner
基座　base
级间衬套　interstage bush
极大的　oversized
集油盘　oil scoop
集油罩　oil housing
几何边缘　geometry edge
几何结构　geometry
几何形状　geometry
几何学上地　geometrically
计量器　gauge
技工　mechanic
迹象　indication
加固物　stiffener
加劲杆　stiffener
加仑每分钟　gallons per minute, gpm
加强　stiffen, strengthen
加强带　tie
加强钉　strengthening spike
加强肋　strengthening rib
加速　accelerate
加压水　pressurized water
加注阀　refill valve
夹带　entrap
假定　assume
尖端　tip
坚固的　robust, sturdy
间接联动式　indirect-acting
间隙　clearance, gap
检测门　inspection gate
检查　inspection
检查口　inspection gate
检修　inspection
检修孔　manhole
检修门　access door
减磨环　wearing ring
减去　minus, subtract
减少了的　diminished
减速　decelerate, retard
减速齿轮　reduction gearing
减速装置　reduction gearing
减压阀　relief valve
剪断销　shear pin
剪切应力　shear stress
简单　simplicity
渐进地　progressively
溅出　spill
降低　degrade
交叉点　intersection
交流发电机　alternator
浇铸　cast
角频率　angular frequency
角速度　angular velocity
接力器　servomotor

接力器活塞　servomotor piston
节距　pitch
节流阀　throttle valve
节流控制　throttle control
节流曲线　throttling curve
节能的　energy efficient
结构　configuration, construction
结构钢　structural steel
结构紧凑　compact structure
结实的　robust
结束　terminate
截流阀　shutoff valve
截止阀　shutoff valve
解释　interpret
紧固件　fastener
紧握　grip
进口部分　inlet section
进口导叶　inlet guide vane
进口段　inlet section
进口法兰　suction flange
进口面积　inlet area
进口直径　inlet diameter
进人孔　manhole
进入　access, admit
进水　incoming water
进水管　distributor

进水角　flow angle
经验的　empirical
精轧的　bright-rolled
井　pit
径流泵　radial flow pump
径流式水轮机　radial flow turbine
径向　radial direction
径向流　radial flow
径向流入式　radial-in-flow
净水头　net head
静排出压头　static discharge head
静水头　static head
静吸入压头　static suction head
静吸上高度　static suction lift
静压　static pressure
静压头　static head
静扬程　static head
矩形底座　rectangular box
矩形盒子　rectangular box
具备　serve
具有破坏性的　destructive
聚四氟乙烯纤维　teflon type fiber
绝对速度　absolute velocity
绝对压力　absolute pressure
均等　even out

K

卡普兰水轮机　Kaplan turbine
卡住的　stuck
开度　opening
开口　opening
开始做…　set about
开挖　excavation
抗力矩　resisting torque
抗磨环　wearing ring
抗转矩　resisting torque
可拆开的　detachable

可动叶片　movable vane
可逆式水泵水轮机　reversible pump turbine
可逆式水力机械　reversible hydraulic machinery
可逆性　reversibility
可旋转的　swivelable
可移动的　movable
克服　overcome
坑　pit
空的　hollow

空化　cavitation
空化特性　cavitational characteristic
空化性能　cavitation performance
空气动力　aerodynamic force
空气动力损失　aerodynamic loss
空气动力学　aerodynamics
空气动力中心　aerodynamic center
空气阀　air valve
空腔　cavity
空转　racing
孔口　orifice
控制　regulate, dictate

控制环　regulating ring, distributor ring
控制机制　controlling mechanism
控制体积　control volume
扣件　fastener
跨度　span
快速旋转　spin
框架　framework
溃灭　collapse
扩散段　diverging section
扩散器　diffuser
扩张　expand

L

拉杆螺栓　stay bolt
拉应力　tensile stress
喇叭形　flared
来水　incoming water
莱斯特·阿伦·佩尔顿　Lester A. Pelton
拦截　intercept
肋拱　rib
类似　resemble
类似的　analogous
累积效应　cumulative effect
冷却　cooling
离心泵　centrifugal pump
离心力　centrifugal force
离心式压缩机　centrifugal compressor
理论基础　theoretical foundation
理论曲线　theoretical curve
理论扬程　theoretical head
立方　cube
立式泵　vertical pump
立轴　vertical axis, vertical shaft
励磁机　exciter
利用　utilize, harness
例证　illustration
连杆　connecting rod

连杆和转臂　links and levers
连接板　connection plate
连接环　link ring
联杆器　crosshead
联轴器　coupler
量　volume
量尺　measurement scale
临界值　critical value
流程　flow path
流出　drain
流道　water passage
流动路径　flow path
流动面积　flow area
流动摩擦损失　flow friction loss
流动逆转　flow inversion
流量　flow rate, capacity, discharge, discharge rate
流速　flow velocity, velocity
流体　fluid
流体动力学　fluid dynamics
流体粒子　fluid particle
流体质点　fluid particle
流线型　streamline
流线型的　streamlined

轮毂　hub, boss, wheel
轮廓　contour
轮流地　in turn
螺杆　screw
螺杆泵　screw pump
螺距　pitch
螺栓　bolt

螺旋撑条　screw stay
螺旋桨　propeller
螺旋桨（式）水轮机　propeller turbine
螺旋千斤顶　jacking screw
螺旋形　volute
螺旋状　spiral-shaped
螺柱　stud

M

马力　horsepower
埋入　embed
埋深　setting depth
脉动　pulsation
锚固螺栓　anchor bolt
弥补　offset
迷宫环　labyrinth ring
米制的　metric
密封　sealing
密封表面　seal surface
密封的　airtight
密封环　sealing ring
密封接头　seal joint
密封面　seal surface
密封圈　sealing ring
描述　depict

敏感性　susceptibility
名义上的　nominal
名义效率　nominal efficiency
铭牌　nameplate
摩擦　friction, rub
摩擦接头　friction joint
摩擦损失　friction loss, frictional loss
摩擦头　friction head
摩擦水头　friction head
磨蚀物质　abrasive material
磨碎　grind
磨损　wear
磨损环　wear ring
模型测试　model test
末端　extremity

N

耐磨环　wear ring
耐磨盘　wear disk
耐用的　robust
挠曲变形　deflection
挠曲应力　flexural stress
内爆　implode, implosion
内表面　inner face
内层　lining
内径　internal diameter
内控式喷射机构　straight flow type injector
内置　build-in

能量　power, energy
能量方程　energy equation
能量损失　loss of energy
能量转换　energy conversion, power conversion
尼龙　nylon
逆时针　anticlockwise
年产量　annual production
碾碎　grind
啮合　mesh
牛顿第二定律　Newton's second law

牛顿定律　Newton's Law
扭矩　torque
扭力销　torque pin

扭曲的　twisted
扭曲应力　torsional stress

O

欧拉　Euler
欧拉方程　Euler's equation

耦合　couple, coupling

P

排出　expel
排出冲程　discharge stroke
排出口　discharge port
排出流　discharge flow
排除　obviate
排掉　drain
排放　discharge
排气阀　air-bleed valve
排气装置　vent fitting
排水　drainage, dewater
排水管线　discharge line
排水配件　drain fitting
旁路控制　bypass control
旁路循环　bypass circulation
旁通阀　bypass valve
旁通管　bypass
抛物线　parabola
刨成（或削成）所需尺寸　dimension
佩尔顿水轮机　Pelton turbine
佩尔顿式水轮机　Pelton wheel turbine
配合　in coordination with
配衡　counter balance
配件　fitting, accessory
配水管　distribution pipe, distributor
配水环　guide ring
配水支管　distribution branch
配送　delivery
配置　configuration
喷杆　spear stem, spear

喷管　injector
喷射机构　injector
喷射力　jet force
喷针　needle
喷嘴　nozzle
偏离　deviation
偏流器　deflector
偏向　deflection
偏斜　deflect
偏转　deflection
拼合式转轮　split runner
频率　frequency
平方　square
平方根　square root
平行　parallel
平行的　parallel
平衡　counter balance, equilibrium
平衡方程　equilibrium equation
平衡孔　balance hole, balancing hole
平衡块　balancing block
平衡装置　balancing device
平炉钢　Siemens Martin steel
平面　plane
平坦的　even
平坦的效率曲线　flat efficiency curve
破坏　breakage
破裂　burst
剖分式蜗壳泵　split volute pump

Q

期初投资　initial investment
鳍形　fin-shaped
起伏　fluctuation
气蚀余量　Net Positive Suction Head, NPSH
气压　air pressure
气泡　vapor bubble
汽轮机　steam turbine
千分之一的　thousandths
千瓦　kW, kilowatt
千瓦特　kW, kilowatt
前缘　leading edge
前锥体　nose cone
嵌入　embed
腔密封　cavity seal
桥式吊车　overhead crane
切断　cut off
切击式水轮机　Pelton turbine, Pelton wheel
切线　tangent
切线的　tangential
切线速度　tangential velocity
切向流式水轮机　tangential flow turbine
切向速度　tangential velocity
切削　trimming
切削的叶轮　trimmed impeller
侵蚀　erosion
倾角　angle of inclination, inclination
倾向于　slant

倾斜　tilt
倾斜　incline, slant
倾斜的部分　oblique division
倾斜地　obliquely
倾斜度　steepness
球面轴承　spherical bearing
驱动　drive
驱动电动机　drive motor
驱动功率　drive horsepower
驱动轮　driving wheel
驱动器　drive
驱动轴　drive shaft
驱动转矩　driving torque
驱逐　expel
屈服　yield
曲柄连杆　toggle lever
曲率　curvature
曲折密封圈　labyrinth ring
曲轴　crankshaft
取代　displace
取得……的专利权　patent
取水（口）　intake
全贯流式水轮机　fully tubular turbine
缺点　cons
缺口　gap
缺少的　absent
确定……的方位　oriented

R

燃料　fuel
燃气轮机　gas turbine
绕过　bypass
热能　heat energy
人孔　manhole
人孔门　access door
人字形的　herringbone

扔弃　dump
容积的　volumetric
容积式泵　positive displacement pump
容积效率　volumetric efficiency
容量　capacity
容纳　accommodate
容器　container

入口　access
入口回旋　inlet rotation
入口井　access shaft
入口旋转　inlet rotation
软管　hose

锐角　acute angle
润滑　lubrication
润滑剂　lubricant
润滑油　lubricating oil, lubricant

S

三脚架　spider
三螺旋高螺距螺旋泵　three-screw, high-pitch, screw pump
闪蒸　flash
上冠　crown
上行程　upstroke
上行冲程　upstroke
上升　rise
上游面　upstream face
上游水位　headwater
上游源头　headwater
勺形　spoon-shaped
少量的液体　slug
设备使用率　plant factor
设计参数　design parameter
射流　jet
伸出　project
伸缩管式连接　telescope connection
伸缩节　telescope connection
伸长　elongation
深入　project into
渗出　seep out
渗漏　leakage
渗漏　leak
渗漏损失　leakage loss
渗漏物　leakage
渗漏油　leakage oil
渗透　penetration
生产　yield
失控工况　runaway condition
失去控制的　runaway

施加（影响、压力等）　exert
十字头　cross head
时时刻刻　from instant to instant
识别　identify
实行　execute
实际曲线　real curve
实线　solid line
实验室测试　lab test
矢量加法　vector addition
使……与……平行　feather
使变圆　rounding
使成锥形　taper
使动作　actuate
使用寿命　service life
示意图　schematic view
市政的　municipal
势能　potential energy
适度的　moderate
室　chamber
收集器　accumulator
收缩　contraction
受油器　oil head, oil transfer unit
枢轴　pivot
输出功率　output power
输入功率　input power
输送　deliver
鼠笼式水轮机　squirrel cage turbine
竖井式贯流机组　pit type tubular unit
竖井通道　access shaft
竖轴　vertical-axis
甩负荷　load rejection, rejection of load

甩油环　oil thrower ring
甩油圈　oil thrower ring
双动式　double acting
双动式顶起螺丝　double acting jacking screw
双缸式　duplex
双击式水轮机　crossflow water turbine, Bánki-Michell turbine, Michell-Banki turbine, Ossberger turbine
双螺旋低螺距螺旋泵　two-screw, low-pitch, screw pump
双曲面　hyperboloid
双曲线体　hyperboloid
双蜗壳泵　double volute pump
双吸　double-suction
双吸式叶轮　double-suction impeller
双重控制　dual control
双重调节　double-regulated
双作用式　double-acting
水泵水轮机　pump turbine
水泵性能曲线　pump characteristic curve
水车　water wheel
水冲力　jet force
水电能源　hydroelectric energy
水斗式水轮机　Pelton turbine
水封的机壳　water tight casing
水封环　lantern ring
水功率　water horsepower
水管　water conduit
水力功率　hydraulic horsepower
水力损失　hydraulic loss
水力效率　hydraulic efficiency
水龙头　tap

水轮机　hydraulic turbine, water wheel, water turbine
水轮机安装高程　turbine setting
水轮机室　turbine chamber
水轮机调速　turbine governing
水能　hydraulic energy, hydraulic power, water energy
水平的　horizontal
水平地　horizontally
水平轴　horizontal axis, horizontal shaft
水头　hydraulic head, head
水头损失　head loss
水推力　hydraulic thrust
水银　mercury
水柱　water column
顺时针　clockwise
瞬时负荷条件　instantaneous load condition
伺服阀　servovalve
松弛的　slack
速度　velocity
速度（流速）剖面　velocity profile
速度分量　velocity component
速度能　velocity energy
速度三角形　velocity triangle
速度矢量　velocity vector
速度梯度　velocity gradient
速度头　velocity head
酸性水　acidy water
随后的　succeeding
碎片　debris
损失　loss
锁定销　locking pin
锁紧圈　clamp ring

T

碳钢　carbon steel
碳氢化合物　hydrocarbon
碳酸水　soda

膛　chamber
套管　sleeve
套环　lantern ring

套筒　sleeve
特性曲线　characteristic curve
梯形的　trapezoidal
提取　extract
提升　lift
提升测试　lift test
提升理论　lift theory
体积　volume
体积的　volumetric
体积流量　volumetric flow rate
填充材料　packing material
填充物　packing
填料　packing
填料函　stuffing box
填料盒　stuffing box
填料环　packing ring
调节　regulate, govern
调节阀　regulation valve
调节杆　regulating lever
调节机构　governing mechanism
调节螺母　adjusting nut
调节器　regulator
调节叶片　adjustable-blade
调节装置　regulating unit, governing device, regulator
调控机构　regulator
调速　governing
调速泵　speed-controlled pump
调速齿轮　timing gear
调速机构　governing mechanism, governor mechanism
调速控制　speed control
调速器　governor
调心滚子轴承　self-aligning roller bearing
调整　adjust

调整杆　regulating lever
调整环　adjustment ring
停机　shutdown
停止位置　idle position
通道　passage
通道门　access door
通风的　vented
同步转速　synchronous speed
同时地　simultaneously
同一地　identically
同源的　homologically
同中心的　concentric
同轴的　coaxial, concentric
同轴地　coaxially
投影　project
投影速度　projection velocity
投资费用　capital cost
透平机械　turbomachine
透平机械　turbomachine
凸轮　cam
凸轮盘　cam disk
突出　protrude
土戈尔式水轮机　Turgo turbine
土建成本　civil cost
土建工程　civil work
湍流的　turbulent
推导出　derive
推导组合轴承　thrust-cum-guide bearing
推拉杆　push-and-pull rod
推力轴承　thrust bearing
托架　bracket
托架轴承　pedestal bearing
托马空化系数　Thoma cavitation coefficient
椭圆形　oval

W

挖掘　excavation

外径　external diameter

外面　outer face
外围　periphery
外形　contour
外缘　outer edge
外罩　housing
弯管　bend
弯曲　deflect
完全相同的　identical
完整性　integrity
往复泵　reciprocating pump
往复活塞　reciprocating piston
微观的　microscopic
微小的　microscopic
围绕　enclose
维护　maintenance
维克多·卡普兰　Viktor Kaplan
尾水　tailwater
尾水管　draft tube
尾水渠　tail race, tailrace, tailrace channel
尾水位　tail water level, tailwater level
尾水锥　draft tube cone

温度传感器　temperature sensor
稳定工况　steady condition
涡流　eddy current, vortex
涡轮　turbo
涡轮机　turbomachine
涡形　volute
蜗壳　scroll case, scroll casing, spiral casing, volute, volute casing
蜗壳舌部　volute tongue
蜗形机壳　volute casing
卧式泵　horizontal pump
卧轴　horizontal axis, horizontal shaft
污水处理厂　sewage treatment plant
污水坑　sump
无摩擦的　frictionless
无人值守的　unattended
无损失理论扬程　theoretical loss-free head
无限的　infinite
无旋的　non-rotational
无叶片的　vaneless

X

吸力面　suction surface
吸入冲程　suction stroke
吸入口　suction port
吸入流　suction flow
吸入室　suction chamber
吸入水头　suction head
吸入压力　suction pressure
吸升水头　suction lift
吸收功率　absorbed power
吸水池　suction tank
吸水地形高度（负）　static suction head
吸水地形高度（正）　static suction lift
吸水高度（负）　suction head
吸水高度（正）　suction lift
吸水管　suction connection

吸水管系统　suction piping
吸水管线　suction line
吸水口　suction nozzle
系统特性曲线　system characteristic curve
下行程　downstroke
下行冲程　downstroke
下环　band
下降　descend
先前的　previous
弦线　chord line
显示　reveal
显著的　salient
现场　site
现场制作　field fabrication
线速度　linear velocity

相当可观的　appreciable
相等的　equivalent
相等地　identically
相对速度　relative velocity
相关　correlation
相互啮合的　intermeshing
相似定律　affinity law
向……提供　feed
向量加法　vector addition
向内破裂　implode
向前的　onward
象征　indication
像　resemble
橡胶圈　rubber ring
消能器　energy dissipater
消散　dissipate
效率　efficiency
效率曲线　efficiency curve
楔块　wedge
斜的　diagonal
斜流式水轮机　Deriaz turbine, Turgo turbine, diagonal turbine
斜坡　slope
谐振　resonant vibration
泄水部件　water drainage part

泄水锥　nose cone
形成空洞　cavitate
性能　performance
性能曲线　performance curve
修改　modification
绪论　exordium
虚线　dashed line
需水量　demand of water
宣布的　declared
悬臂　cantilever
悬臂式的　overhung
旋桨泵　propeller pump
旋转　rotate, revolve, whirl
旋转泵　rotary pump
旋转分量　whirl component, whirling component
旋转机械　rotating machinery
旋转运动　rotary action
旋转轴　axis of rotation, rotating shaft, rotation axis
旋转组件　whirl component
漩涡　eddy, vortex
削尖　sharpen
循环　circulate
循环泵　reciprocating pump

Y

压盖　gland
压盖填料　gland packing
压降　pressure drop
压力　stress
压力计　pressure gauge
压力面　pressure surface
压力容器　pressure vessel
压力水管　penstock
压力水头　pressure head
压力梯度　pressure gradient
压力下降　pressure drop

压力作用　pressure action
压能　pressure energy
压水池　discharge tank
压水地形高度　static discharge head
压水管　discharge connection
压缩机　compressor
压头　pressure head
延迟　retard
延长　elongation
严重的　severe
扬程　head

扬程-流量特性曲线 head-discharge characteristic curve
扬程曲线 head curve
扬程损失 head loss
叶槽 impeller passage
叶道 blade channel
叶轮 impeller
叶轮流道 impeller passage
叶轮切削 impeller trimming
叶轮入口 impeller eye
叶轮直径 impeller diameter
叶片 blade, bucket, vane
叶片安放角 blade angle
叶片背面 suction side, suction surface
叶片臂 blade arm
叶片操作机构 blade-operating mechanism
叶片负压侧 suction side
叶片工作面 pressure side
叶片形状 blade shape
叶片正面 pressure side, pressure surface
叶片轴 blade axis
叶片转臂 blade lever
叶栅 blade row
液压动力 hydraulic power
液压推力 hydraulic thrust
液压压头 hydraulic head
依次 in turn
依附 adhere
移除 remove
异常的 exceptional
异物 foreign object
异形 profiled
异型钢 profiled steel
易受影响或损害的状态 susceptibility
溢出 spill
翼形 aerofoil shaped
翼形理论 airfoil theory
翼型 airfoil
翼型叶片 airfoil blade

翼状的 wing-shaped
引发 trigger, incur
引水部件 water diversion part
引水道 guide passage
英尺 feet
英寸 inch
英制 English system
英制单位 English unit
迎角 angle of attack
迎水面 leading face, pressure side
应力 stress
硬化的 hardened
永态转差率 permanent droop
用螺栓固定 bolt
优点 pros, virtue
油管 oil pipe, oil piping
油管吊索 oil sling
油膜 oil layer
油位浮标 level float
油位指示器 oil level indicator
油箱 oil tank
油压系统 oil pressure system
油脂 grease
有抵抗力的 resistant
有腐蚀性 aggressive
有规律地跳动 pulsate
有限的 finite
有效水头 available head, effective head
右旋螺纹 right-handed thread
淤泥 silt
余量 allowance
余弦 cosine
与……协调 in coordination with
预应力 prestressed
原点 origin
原动机 prime mover
原水 raw water
圆弧 arc
圆盘摩擦 disk friction

圆盘摩擦损失　disk friction loss
圆头铁锤　ball pein hammer
圆周　circumference, periphery
圆柱体　cylinder
圆柱体的　cylindrical
圆柱形的轮辐　cylindrical disc
圆柱形转轮　cylinder-shaped runner
圆锥的　conical
允许工作范围　allowable operating range, preferred operating range
允许气蚀余量　Net Positive Suction Head Available
运动部件　operating member
运营成本　operating cost

Z

载荷限制器　load limiter
再分　subdivided
在……之前　precede
噪音水平　noise level
增加　gain
黏度　viscosity
黏附　adhesion
黏性流体　viscous fluid
詹姆斯·比切诺·法兰西斯　James B. Francis
占优势的　dominant, prevailing
招致　incur
折向器　deflector
着手　set about
针阀　spear rod
真空　vacuum
蒸发　vaporization, vaporize
蒸汽活塞　steam piston
蒸汽轮机　steam turbine
蒸汽压　vapor pressure
整周进水式水轮机　full admission turbine
正齿轮　spur gear
正功　positive work
正交的　perpendicular
正面　leading face
正排量泵　positive displacement pump
正切　tangent
支臂　arm
支撑环　support ring
支撑连杆　link support
支承轴承　pedestal bearing
支架　bracket, spider
支配　dictate
执行　execute
执行机构　actuator
直角　right angle
直角转弯　right-angle turn
直接联动式　direct-acting
直流式喷射机构　straight flow type injector
直喷嘴　straight flow type injector
止回阀　check valve, non-return valve
止推轴承　thrust bearing
止推轴承　axial bearing
指导方针　guideline
指定　designate
指针　pointer
制动功率　brake horsepower
制动射流　braking jet
制造　fabricate
制造商　manufacturer
制止　check
质量流量　mass flow, mass flow rate
滞止焓　stagnation enthalpy
中等的　moderate
中腹部　midriff
中间齿轮　idler
中间的　intermediate
中间过渡段　intermediate transitory section
中间接力器　pilot servomotor

中空的 hollow	逐步的 progressively
中跨 mid-span	主电路开关 main circuit switch
中心线 centerline	主接力器 main servomotor
终止 terminate	主涡 primary vortex
种类 category	主支柱 main stay
重叠 overlap	主轴 main shaft
重叠的孔 overlapping bore	主轴密封 shaft seal
重力 gravitation	煮沸 boil
重新开始 resume	注入 feed, inject
轴 axis, shaft, spindle	注射 inject
轴承 bearing	注油孔 filling opening
轴承垫片 bearing pad	柱塞 plunger
轴承损失 bearing loss	柱形阀 spool valve
轴承箱 bearing housing	铸钢 cast steel
轴承座 bearing housing	铸轮 cast wheel
轴封 shaft seal	专利权 patent
轴封盒 shaft seal box	转臂 steering arm, lever
轴封损失 shaft seal loss	转动部分 rotating element
轴封装置 shaft seal, shaft seal device	转动元件 rotating element
轴功率 shaft power	转动轴 axis of rotation, rotation axis
轴流 axial flow	转换 convert
轴流泵 axial flow pump, propeller pump	转桨式水轮机 Kaplan turbine
轴流定桨式水轮机 fixed blade propeller turbine	转矩销 torque pin
轴流式水轮机 axial flow turbine	转轮 runner, rotor, wheel
轴流转桨式水轮机 adjustable blade Kaplan turbine	转轮室 runner chamber
	转轮叶片 runner vane
轴伸式贯流机组 shaft extension type tubular unit	转每分钟 revolutions per minute
	转速 rotational speed
轴伸式水轮机 S-type turbine	转向 deflect
轴套 axle sleeve	转移 divert
轴瓦 bearing bush	转轴 rotating shaft
轴线 axial line	转子 rotor
轴向 axial direction	转子泵 rotary pump
轴向力 axial force	装配 assembly
轴向流动 axial flow	装置 installation
轴向推力 axial thrust	装置角 setting angle
轴向隙角 axial relief	撞击 impact, impinge
肘形 elbow shape, elbow-shaped	锥管段 conical section
	锥形轮毂 conical hub

资本成本	capital cost
子类型	subtype
自动排气阀	automatic air valve
自灌式	self-priming
自吸式	self-priming
总动排出压头	total dynamic discharge head
总动吸入高度	total dynamic suction lift
总动吸入压头	total dynamic suction head
总动吸上高度	total dynamic suction lift
总动压头	total dynamic head
总静水头	total static head
总水头	total head
总体效率	overall efficiency
总效率	total efficiency
总压头	total head
纵轴	vertical axis
纵坐标	ordinate
阻碍	clog
阻力	resistance
阻力矩	resisting torque
阻塞的	clogged
组装	assembly
最初投资	initial investment
最大效率	maximum efficiency
最高效率	peak efficiency
最适宜的	optimum
最优效率	optimum efficiency
遵照	conform
左旋螺纹	left-handed thread
作业时间	operating time
座环	stay ring, stay cone